PROGRESS SERIES

ORGANIC CHEMISTRY

7

PROGRESS IN
ORGANIC CHEMISTRY
7

Joint Editors
Sir JAMES COOK, D.Sc., LL.D., F.R.S.
Vice-Chancellor
University of East Africa
and Fellow of University College
London

and

W. CARRUTHERS, Ph.D.
Lecturer in the Department of Chemistry
University of Exeter

SPRINGER SCIENCE+BUSINESS MEDIA, LLC

First published by
Butterworth & Co. (Publishers) Ltd.

ISBN 978-1-4899-7299-6   ISBN 978-1-4899-7315-3 (eBook0
DOI 10.1007/978-1-4899-7315-3

© Springer Science+Business Media New York   1968

Originally published by Butterworth & Co. (Publishers) Ltd.in 1968
Softcover reprint of the hardcover 1st edition 1968

Suggested U.D.C. No: 574 (047.1)
Library of Congress Catalog Card Number 52–3180

FOREWORD

FOLLOWING the pattern of earlier volumes of the series, five themes covering a range of topics of current interest to organic chemists are discussed in the present volume. Two of the chapters are concerned directly with the chemistry of natural products and one with a reaction which is of importance in a number of fundamental biochemical processes. A fourth chapter reviews an interesting series of rearrangement reactions and the remaining chapter is concerned with an area of physico-organic chemistry. As with earlier volumes, the authors are all specialists who have themselves contributed to the fields of work which they have reviewed.

The tetracycline antibiotics have fascinated organic chemists since the discovery of the first member of the series in 1947. As a result of sustained research the chemistry of the complex array of functional groups on the tetracyclic framework is now approaching a stage comparable to that found in the steroid series, where a variety of interesting and selective chemical changes can be induced at different positions in the molecule. The biosynthesis of the tetracyclines has also provided a problem of deep interest, now well on the way to solution, and the formidable challenge of total synthesis has been accepted in several laboratories and met to a degree which might not have seemed possible at the outset. All of these aspects are touched on by Dr. Money and Professor Scott in their illuminating account of the chemistry and biochemistry of the tetracycline antibiotics.

In the second chapter, Dr. Habermehl reviews the interesting class of the salamander alkaloids. It has been known for a long time that the black and yellow spotted fire salamander is venomous and that the skin gland secretion is the source of the toxicity. Recent investigations have shown that the toxic material is a mixture of closely related basic substances containing steroid-like skeletons with a modified ring A. X-ray crystallographic analysis played a large part in the elucidation of the finer points of the structure of these compounds.

The third chapter is concerned with electrophilic molecular rearrangements. In his lucid survey Professor Stevens gives a very full account of the variety of forms which rearrangements of this kind can take and of the structural features which favour them. Several rearrangements in this series have useful synthetic applications.

The importance of phosphoryl transfer reactions in a number of fundamental biochemical processes is now well appreciated, and recent studies in the laboratory have thrown much light on the pathway by which these reactions may be effected in Nature. A considerable range of chemical phosphorylating agents has been discovered, and Professor Clark and Dr. Hutchinson provide a valuable survey of the different types and of the conditions under which phosphorylation may be effected, emphasizing the biochemical implications of the different methods. Recent detailed work on the biological phosphorylation of adenosine diphosphate suggests that it may involve reactions which closely parallel some which have been successfully accomplished *in vitro*.

In the last chapter, Dr. Fischer and Dr. Rewicki consider the determination of acid strengths of acidic hydrocarbons from both the theoretical and practical points of view. These acids are nearly always π electron systems and this survey records progress which has been made in the theoretical evaluation of acid strengths by application of quantum theory. The synthesis and reactions of the hydrocarbon acids and of the related cyanocarbon acids are also discussed.

J. W. COOK
W. CARRUTHERS

CONTENTS

		PAGE
FOREWORD		*v*

1 RECENT ADVANCES IN THE CHEMISTRY AND BIOCHEMISTRY OF TETRACYCLINES   1

    T. MONEY, Ph.D., Lecturer in Chemistry, University of Sussex, Brighton

    A. I. SCOTT, Ph.D., D.Sc. Professor, Chemical Laboratory, University of Sussex, Brighton

2 SALAMANDER ALKALOIDS   35

    G. HABERMEHL, Priv. Dozent Dr., Institut für organische Chemie, der Technischen Hochschule, Darmstadt, W. Germany

3 ELECTROPHILIC MOLECULAR REARRANGEMENTS   48

    T. S. STEVENS, D.Phil., F.R.S., Emeritus Professor, Department of Chemistry, University of Sheffield

4 PHOSPHORYL TRANSFER   75

    V. M. CLARK, M.A., Ph.D., Professor, School of Molecular Sciences, University of Warwick, Coventry

    D. W. HUTCHINSON, Ph.D., A.R.I.C., Lecturer, School of Molecular Sciences, University of Warwick, Coventry

5 ACIDIC HYDROCARBONS   116

    H. FISCHER, Dr.rer.nat., Lecturer in Chemistry, Max-Planck-Institut für Medizinische Forschung, Heidelberg, W. Germany

    D. REWICKI, Dr.rer.nat., Institut für organische Chemie der Freien Universität, Berlin, W. Germany

INDEX

1

RECENT ADVANCES IN THE CHEMISTRY AND BIOCHEMISTRY OF TETRACYCLINES*

T. Money and A. I. Scott

INTRODUCTION	1
TETRACYCLINES OF NATURAL ORIGIN	2
CHEMICAL REACTIVITY	6
BIOSYNTHESIS OF TETRACYCLINES	18
TOTAL SYNTHESIS OF TETRACYCLINES	24
The Braunschweig–Madison Syntheses	26
The Pfizer–Harvard Synthesis	27
The Lederle Synthesis	29
The Moscow Synthesis	29
CONCLUSION	30

SINCE 1947, when *aureomycin*[1], the first member of the family of tetracycline antibiotics was described, there has been sustained an ever-deepening interest in the chemistry of these important therapeutic agents[2]. Several reviews emphasizing the chemistry[3-7], synthesis[6, 8, 9] and biological activity [7, 10, 11] of the tetracyclines have been published. It has become apparent that the chemistry of the complex array of functions present on the linear tetracyclic framework is now approaching a stage comparable to that of steroid chemistry, requiring cognizance of selective reaction at each centre of the molecule.

The details of extensive degradative studies which allowed structure (I) to be proposed[12] for *terramycin* (5-hydroxytetracycline) in 1952 clearly showed that many interesting chemical changes could be wrought at several positions in the molecule. A second important aspect of tetracycline chemistry has been the study of the biosynthesis of these acetogenic[13] metabolites. Furthermore, the redoubtable challenge of total synthesis has been accepted in several laboratories and, indeed, met to a degree which might not have seemed remotely possible at the outset, bearing in mind the sensitivity towards degradation encountered in the early chemical studies.

This chapter is concerned with recent advances made in these three

* The literature review for this chapter was completed in May 1965.

areas of endeavour viz. chemical reactivity, biosynthesis and total synthesis of the tetracycline family, prefaced by a short description of the naturally-occurring members.

TETRACYCLINES OF NATURAL ORIGIN

1. *Terramycin* (5-Hydroxytetracycline), *Aureomycin* (7-Chlorotetracycline) and *Tetracycline*—The gross structure (I) for terramycin was deduced from a wealth of experimental data in 1952[12]. At the same time certain stereochemical features could be discerned. Thus, a *cis*-relationship at *4*a, *12*a (dehydration difficult) and *trans*-5a,6 stereochemistry (dehydration facile) were deduced. The relative configurations of the 5-hydroxyl and 4-dimethylamino groups were more difficult to determine, especially in view of the ease of epimerization[14-17] of the latter. However, indirect evidence favoured stereochemistry (II), for

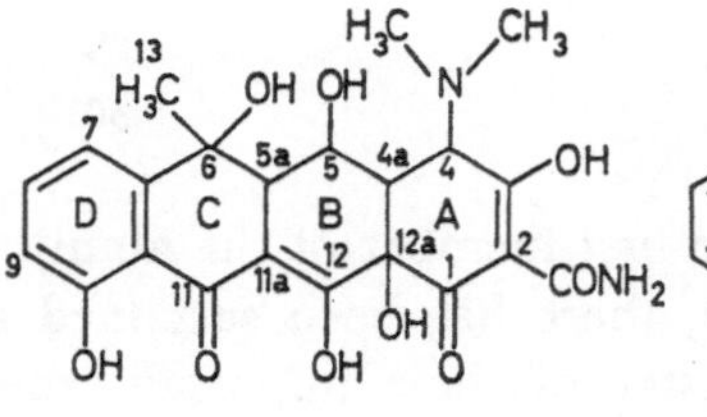

(I)

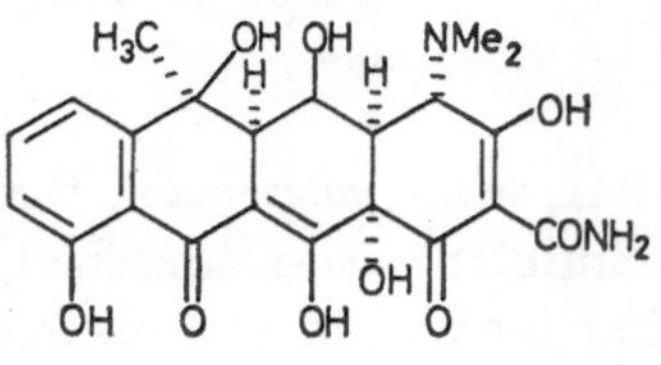

(II)

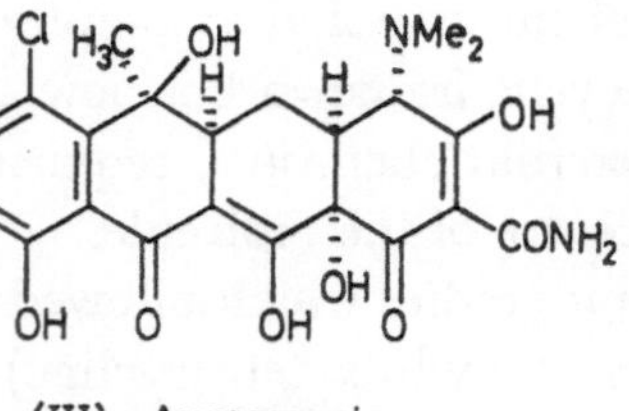

(III) Aureomycin

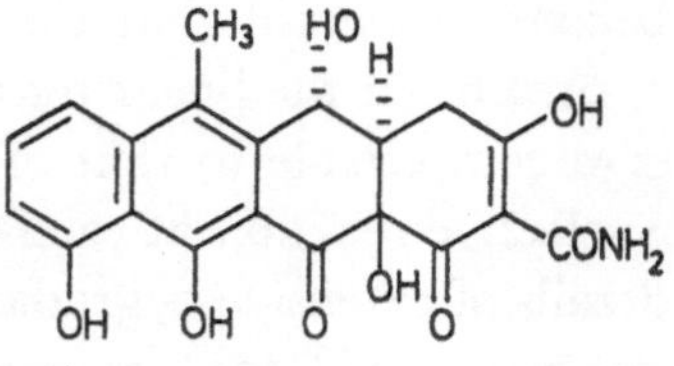

(IV)

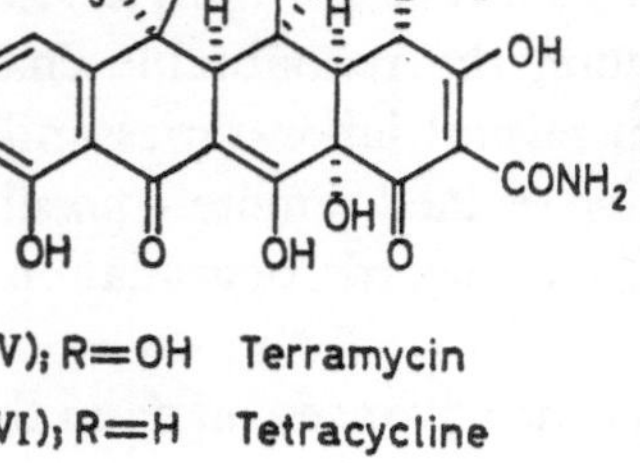

(V); R=OH   Terramycin

(VI); R=H   Tetracycline

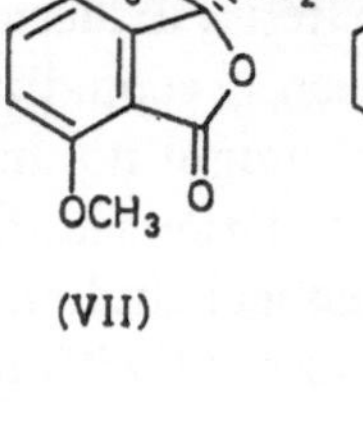

(VII)

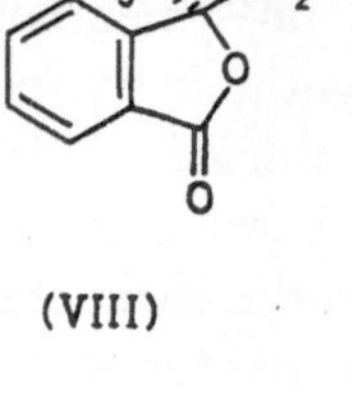

(VIII)

2

certain reactions involving $5 \rightarrow 12$ bridging could be more readily explained on the basis of this, rather than the corresponding *epi* configuration. Thus, whilst the stereochemistry of aureomycin (7-chlorotetracycline) was defined as in (III) by X-ray studies[18], the corresponding diffraction data for terramycin hydrochloride[19] resulted in a view of C_5-stereochemistry along an axis which left considerable ambiguity as to the true configuration at this centre. The 5α-configuration was recently established[20] by taking advantage of the *trans*-stereochemistry at the A/B ring junction in 12a-*epi*-4-desdimethyl-amino-5a,6-anhydro-5-hydroxytetracycline (IV). The nuclear magnetic resonance signals of the C_5 and C_{4a} protons in this compound showed a coupling constant of 8 c/s leading to a *trans*-relationship of these hydrogens and the resultant complete stereochemistry (V) for terra-mycin. Since aureomycin and tetracycline are simply related by chemistry not affecting an asymmetric centre, the complete configuration (VI)

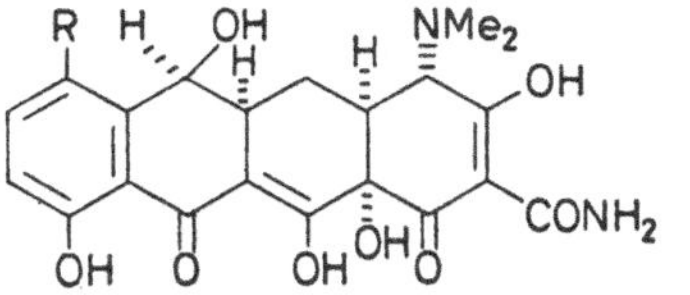

(IX) R=H
(X) R=Cl

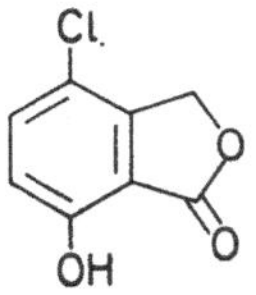

(XI)

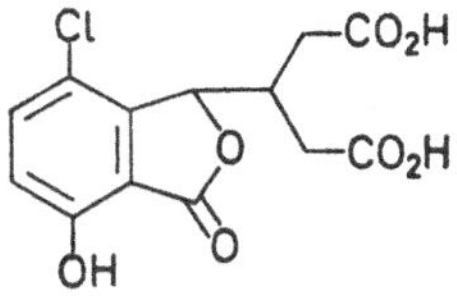

(XII)

may be written for tetracycline, the parent of the series and the third naturally occurring member. The absolute configuration (III) of aureomycin (and by analogy of the other natural tetracyclines) was deduced from the optical rotatory dispersion curve of the degradation product (VII) which mirrored that of (VIII), of proven absolute stereochemistry[20a].

2. *6-Demethyltetracyclines*—Earlier extensive degradation studies with terramycin (V) paved the way for rapid elucidation of the structures of new tetracyclines isolated from various mutant strains of *Streptomyces* species. Thus in 1957 a new strain of *S. aureofaciens* produced 6-demethyltetracycline (IX) and its 7-chloro derivative (X)[21, 22].

Appropriate degradation, including pyrolysis of (X) to the phthalide (XI), demonstrated the absence of the C_6 methyl group whilst formation of 5a,6-anhydro derivatives and alkaline degradation (X) → (XII) determined that the 6-demethyltetracyclines possessed essentially the same structure as the tetracyclines themselves and that the 6-hydroxyl group bears the same configuration in both the methylated and non-methylated series[23].

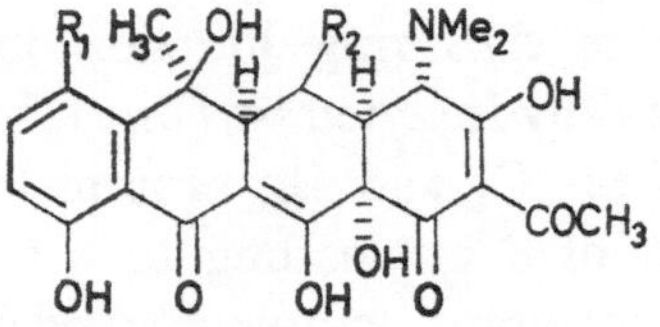

(XIII) $R_1 = R_2 = H$
(XIV) $R_1 = H; R_2 = OH$
(XV) $R_1 = Cl; R_2 = H$

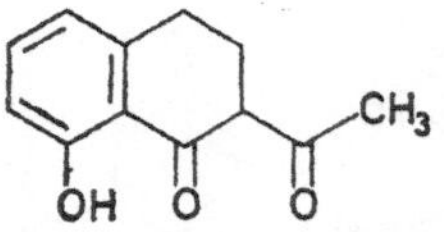

(XVI) 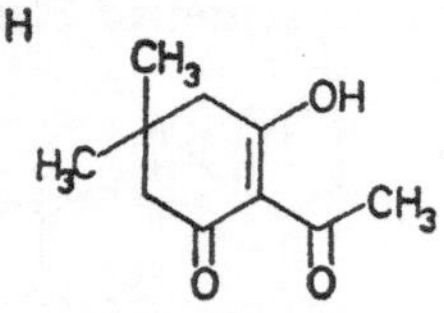   (XVII)

3. *7-Bromotetracycline*—Replacement of chloride with bromide ion in *S. aureofaciens* fermentations leads to the production of 7-bromotetracycline[24](III; Cl = Br).

4. *2-Acetyldecarboxamidotetracyclines*—Evidence (albeit circumstantial) that tetracycline biosynthesis is based on the acetate/malonate pathway can be adduced from inspection of the structures of 2-acetyl-2-decarboxamidotetracycline (XIII)[26] and its 5-hydroxylated (XIV) and 7-chloro derivatives (XV)[25], which are elaborated by mutant strains of *S. aureofaciens* and *S. rimosus*. The importance of ultraviolet spectroscopy, still perhaps the most vital physical method in the classification and analysis of tetracyclines, is illustrated by part of the structure proof for the acetyl tetracyclines. Degradative and comparative experiments showed that (XIII) was quite similar to terramycin. However, no carboxamido group was present in (XIII), and in contrast to the 'normal' tetracyclines, Kuhn–Roth oxidation afforded 2 molecules of acetic acid. The carbonyl stretching frequency in the infra-red spectrum of (XIII) at 1670 cm^{-1} (*cf.* tetracycline with no $>C=O$ absorption above 1665 cm^{-1}) indicated that the —$COCH_3$ side chain should be placed on ring A; any other positioning would have modified the ring-BCD

4

chromophore characteristic of the tetracyclines, and also present in the 2-acetyl series. A final choice in favour of position 2 was made when subtraction of the u.v. spectrum of 2-acetyl-8-hydroxytetralone (XVI) from that of (XIII) gave a curve identical with the spectrum of 2-acetyl-dimedone (XVII).

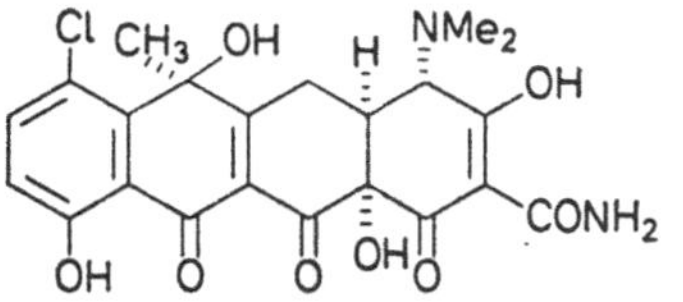

(XVIII) Ketonic tautomer 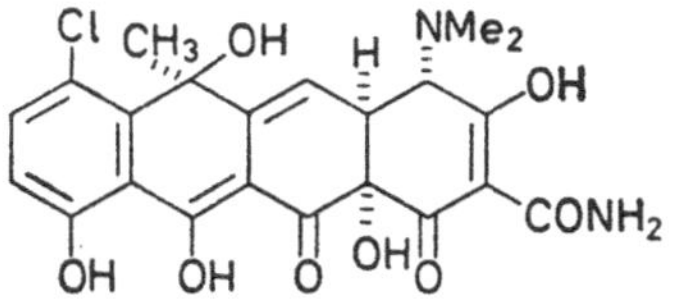   (XIX) Enolic tautomer

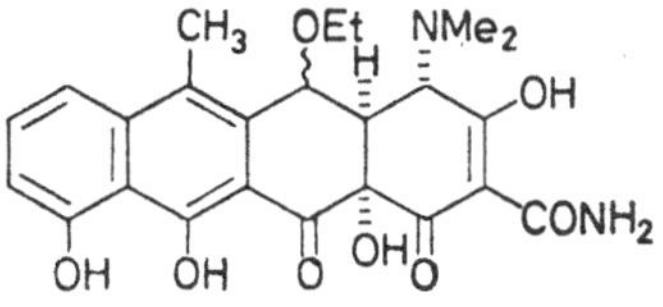

(XX)

5. *'Biosynthetic' Tetracyclines*—Important contributions to the problem of tetracycline biosynthesis have been provided by isolation (principally at the Lederle Laboratories) of modified tetracyclines, several of which are capable of biological conversion to the parent antibiotics. Thus 5a,11a-dehydro-7-chlorotetracycline (XVIII)[27] is accumulated by mutants of *S. aureofaciens* and can be converted into aureomycin by further fermentation (see p. 20). It was possible to isolate two isomeric forms of (XVIII) by recrystallization from different solvents. The $\Delta^{5a,11a}$-isomer (XVIII) (from chloroform) has $\nu(C=0)$ 1716 cm^{-1}, while the $\Delta^{5,5a}$-isomer (XIX) (from water) has no $\nu(C=0)$ stretching frequency above 1660 cm^{-1}. These assignments have found support from n.m.r. studies[28]. Catalytic reduction of (XVIII) affords successively 7-chlorotetracycline and tetracycline together with a considerable proportion of the appropriate 5a-epimer[27] in each case. 5a,11a-dehydro compounds are important not only as relays in biosynthesis but as intermediates for tetracycline synthesis. Furthermore, 5-oxygenated anhydrotetracyclines (as XX) can be prepared from (XVIII) by treatment with alcohols under acidic conditions[28].

More recently representatives of the C-4 modified tetracyclines (XXI; R = Et)[29] and (XXIa; R = H)[30] have been isolated. The chemical and biological conversion of 5a,6-anhydro-4-dedimethylamino-4-amino tetracycline (XXII) to anhydrotetracycline not only corroborated the

5

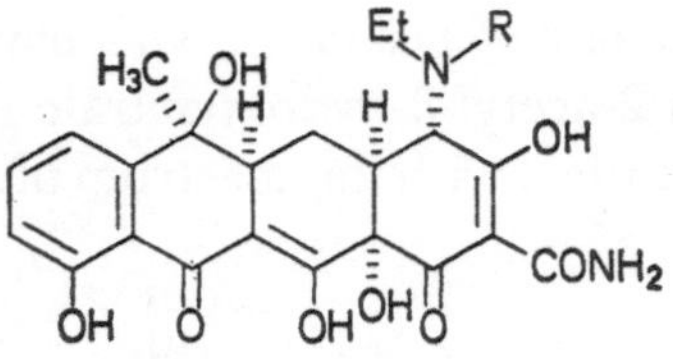

(XXI) 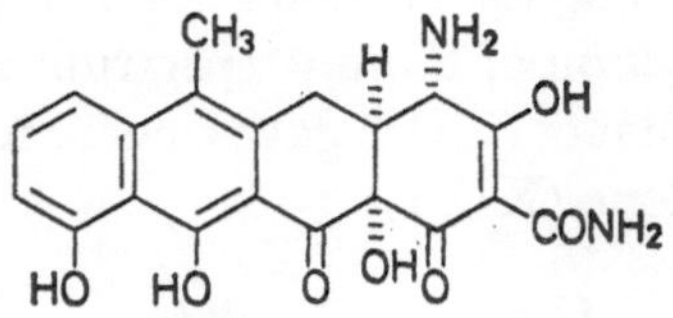 (XXII)

assigned structure but indicated the biosynthetic sequences of N-methylation and anhydro → 5a,11a-dehydro conversion (see p. 21).

CHEMICAL REACTIVITY

C_2—Several N-alkyl derivatives of the 2-carboxamido function have been prepared, e.g. the amino methyl (**XXIII**) and the corresponding

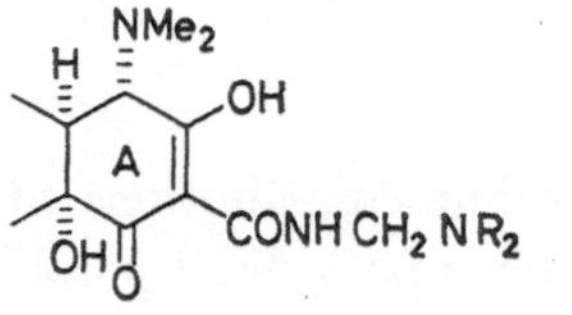

(XXIII) 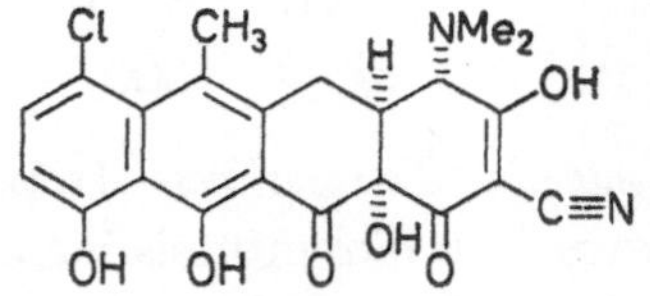 (XXIV)

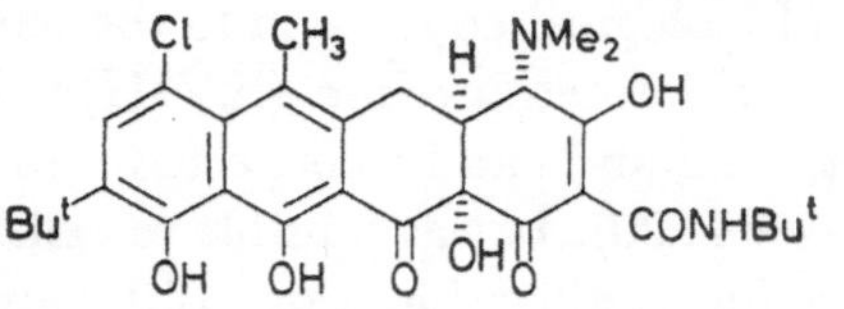

(XXV)

t-butyl compounds[31, 32]. Dehydration of aureomycin at both 5a,6 and the carboxamide grouping using methane sulphonyl chloride in pyridine affords anhydroaureomycin nitrile (**XXIV**) which in turn was converted, by treatment with isobutene-sulphuric acid into (**XXV**)[33].

C_4—(a) *Epimerization*—Early observations of the chemistry of the tetracyclines, as well as pointing to the relative stereochemistry at 5a,6 and 4a,12a, indicated that a reversible epimerization could be brought about at slightly acidic pH. That this change involved C_4 could be demonstrated by conversion of aureomycin and its epimer into the nitriles (**XXVI**) and (**XXVII**) respectively which still retained the epimerizable centre[14–17]. The configuration at C_4 was ultimately settled by X-ray analysis[18].

6

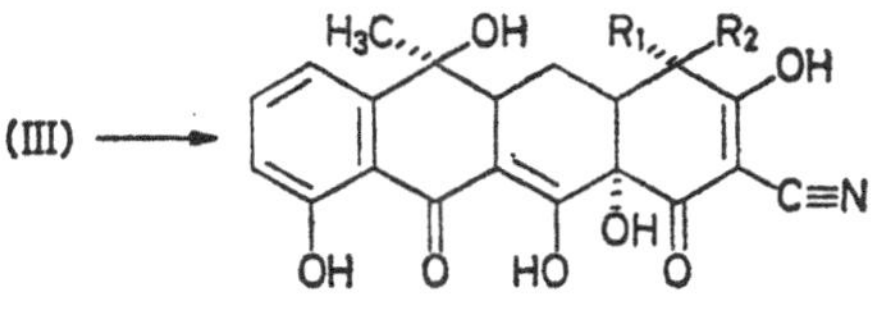

$$(III) \longrightarrow$$

(XVI) $R_1 = NMe_2$; $R_2 = H$
(XVII) $R_1 = H$; $R_2 = NMe_2$

Considerable loss of biological activity accompanies C_4-epimerization and it has therefore become important to apply rigorous control to this equilibrium[34]. The 'normal' configuration is favoured by the use of calcium, magnesium and strontium salts; whereas an equilibrium mixture of 4-epimers is usually obtained at pH 5–7 the addition of calcium ion and adjustment to pH 8–10 results in virtually complete regeneration of the normal series[34].

(b) *Removal of the Dimethylamino Group*—Selective *reductive* removal of the C_4-dimethylamino function is achieved by methylation to the quaternary ammonium iodide followed by brief treatment with zinc and 50 per cent acetic acid (XXVIII)[36, 37]. Under more vigorous conditions, use of the same reagent results in loss of both the 12a-hydroxyl and 4-dimethylamino functions (XXIX).

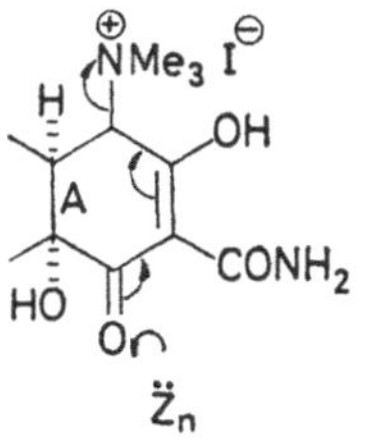

(XXVIII) 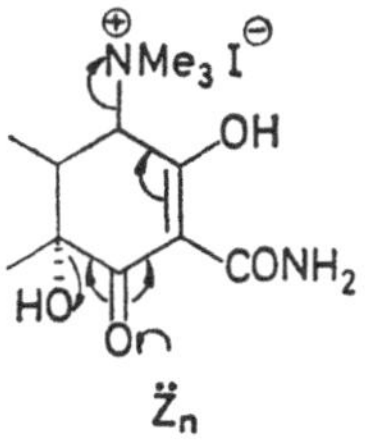   (XXIX)

In the case of terramycin, participation of the C_5-hydroxyl group with the eliminating quaternary salt leads to the bond cleavages depicted in (V) → (XXX) + (XXXI)[35].

Under carefully defined conditions positive halogens, air, cupric and mercuric acetates selectively induce oxidative removal of C_4-nitrogen to generate 4-oxotetracyclines, a reaction reminiscent of the conversion of tertiary amines to aldehydes with hypochlorite ion. The participation of the 6-hydroxl group in this reaction is stereochemically very favourable

7

and the intermediate (XXXII) has been isolated[36]. The 4-oxotetra-
cyclines exist as the 6 → 4 hemiketals e.g. (XXXIII), and the reaction has so far been applied successfully to tetracycline, 7-chlorotetra-

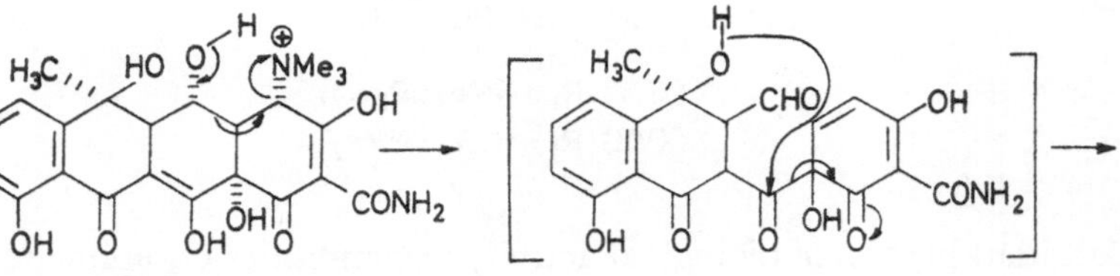

Terramycin (V)

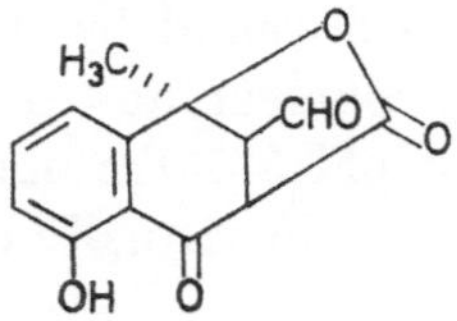

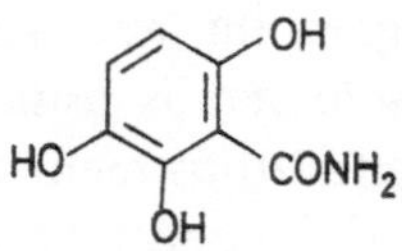

(XXX)   (XXXI)

cycline and their 6-*de*methyl derivatives. Reductive amination of the
tetracycloxide (XXXIV) followed by reductive methylation to the
6-demethyltetracycline not only proved the tetracycloxide structure
but also illustrated the many possibilities for changing the basic centres
at C_4. Among the derivatives prepared in this way are the 4-amino,
4-methylamino, 4-ethylamino, 4-*n*-propylamino set as well as the
methyl ethyl, methyl propyl and diethylamino compounds. The amino
function can be inserted directly by reductive amination[36] or via oxime
or hydrazone formation[37]. In all of these reactions the 4-*epi* configuration is produced.

The correct choice of media for tetracycloxide formation is vital, for
reaction of N-chlorsuccinimide (N.C.S.) in all but aqueous solutions
with tetracycline (VI) affords the 11a-chloro compound (XXXV)
whose (blocked) BCD ring system has an ultraviolet spectrum almost
identical with that of the non-enolizable tetracycloxide, the latter
retaining the 11,12-β-dicarbonyl system in the keto form (XXXVI)[36, 37].
The 11a-chloro compound prepared by N.C.S. in $CHCl_3$ can be further
oxidized with aqueous N.C.S. to the chloro oxide (XXXVII)[37].

C_5—The principal reactions of the hydroxyl group at C_5 were clearly
delineated in the classic paper[12] on the structure of terramycin. Since
that time the main interest has been the definition of C_5-stereochemistry
which has recently been secured[20]. As mentioned above, the principal

8

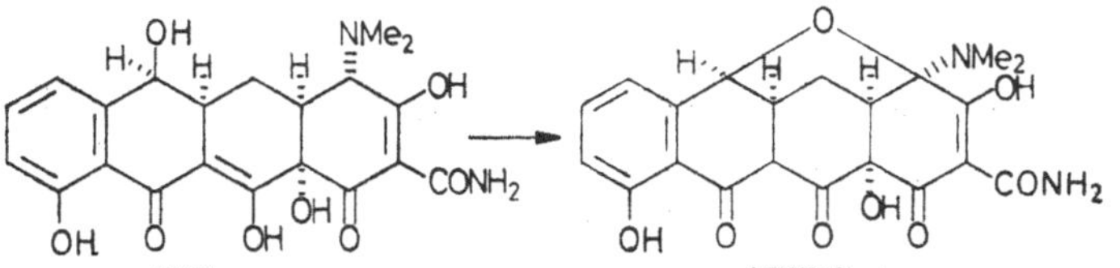

(IX)   (XXXII)

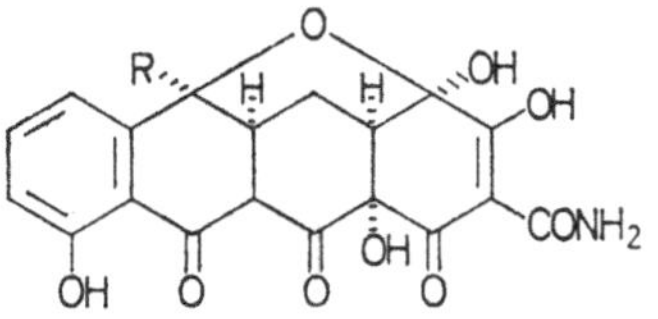

(XXXIII) R = CH₃
(XXXIV) R = H

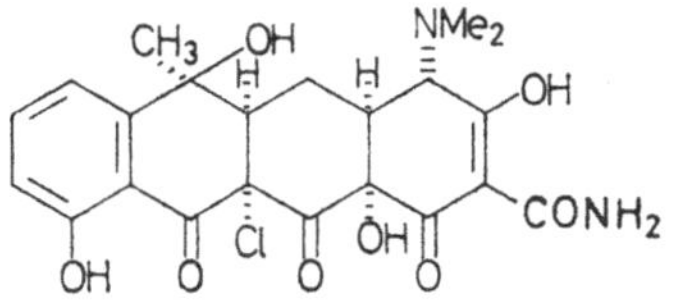

(XXXV)

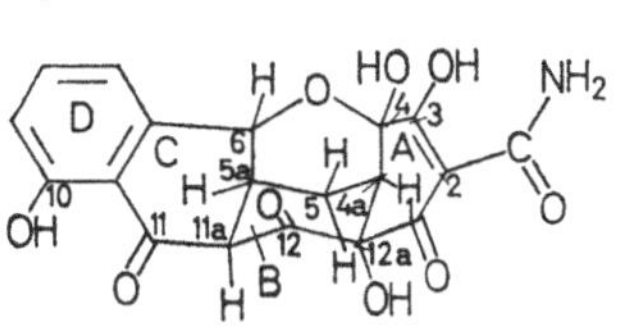

(XXXVI) 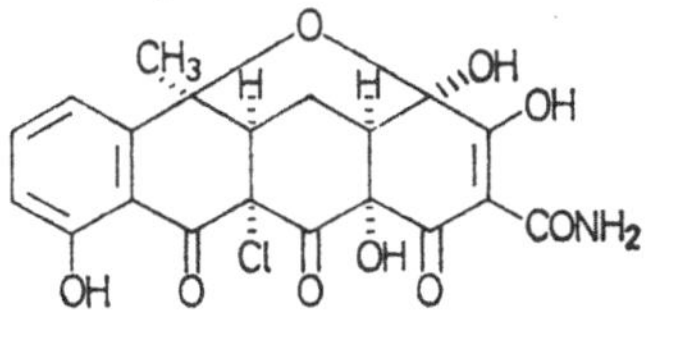   (XXXVII)

chemical study concerned with this centre has been the successful intro-
duction[28] of C₅-alkoxyl groups into 7-chloro-5,5a-dehydrotetracycline
(XVIII) using alcoholic hydrogen chloride solution. Acetylation of the
C₅-hydroxy of terramycin has been reported[40].

C₆—6-*Deoxytetracyclines*—Hydrogenolytic removal of the 6-hydroxyl
function from both the tetracycline (XXXVIIIa → XXXIXa) and
6-demethyltetracycline (XXXVIII → XXXIX) series not only leaves
the biological activity of the appropriate member unimpaired, but
confers sufficient stability on the resultant 6-deoxy compound to allow
electrophilic aromatic substitution to operate in ring D.

Epimerization at C₆ accompanies 6-deoxygenation of tetracycline and
5-hydroxytetracycline, a result which had been anticipated during
extensive synthetic investigations by MUXFELDT[41]. Since a noble metal
catalyst in acidic medium is necessary for C₆ hydrogenolysis, concurrent
5a,6-dehydration is a competing side reaction. A result of importance
for synthetic studies (see p. 24) is the finding that not only does
6-demethyltetracycline (XXXVIII) undergo 6-deoxygenation in 30–40
per cent yield, but the resultant 6-demethyl-6-deoxytetracycline
(XXXIX) shows the full antimicrobial spectrum of the tetracyclines
proper.

Electrophilic substitution of ring D of 6-demethyl-6-deoxytetra-
cycline (XXXIX) proceeds smoothly without disruption of the

(XXXVIII) R = H
(XXXVIIIa) R = CH$_3$

(XXXIX)   R$_1$ = R$_2$ = H
(XXXIXa) R$_1$ = H; R$_2$ = CH$_3$ 'β'
(XXXIXb) R$_1$ = CH$_3$; R$_2$ = H 'α'

molecule. For example, nitration[42] (potassium nitrate–sulphuric acid) affords a mixture of the 7- and 9-nitro derivatives [(XL) and (XLII) R = NO$_2$] convertible in turn to the amino and diazonium compounds by standard methods. It is of interest that the 7-nitro compound (XL; R = NO$_2$) is twice as active *in vitro* against test organisms (*S. aurens* assay) as tetracycline although *in vivo* results were disappointing[7]. Halogenation of 6-deoxy-6-demethyltetracycline (XXXIX) can be directed by acidity control. In experiments with 7-tritiated starting material (XL; R = T) N-bromo and N-iodosuccinimide form the 7-bromo and 7-iodo compounds respectively in sulphuric acid, whereas in acetic acid the 11a-halo-6-demethyl-6-deoxytetracycline (XLI; R = Cl) is isolated (v 1739 cm^{-1}; ring BCD chromophore interrupted)[43]. The dependence of 11a-halogen stability on reaction conditions could be demonstrated by the acid catalysed (concentrated H$_2$SO$_4$) re-arrangement of the 11a-bromo compound (XLI; R = Br) to the 7-bromide (XL; R = Br). Competition experiments using α-naphthol established that the rearrangement is intermolecular.

Nucleophilic substitution of the 7- and 9-diazonium compounds has been observed. Thus azide ion has been used to replace the 7-diazonium group. Reaction of the 9-diazonium sulphate (XLII; R = N$_2$$^+$) with methanol effected reduction back to (XXXIX)[43].

Photolysis of the 7-diazonium sulphate hydrochloride of (XXXIX) in acetic acid solution gave a mixture of 7-chloro and 6-acetoxy-6-de-methyl-6-deoxytetracycline together with (XXXIX)[44]. Photo-decom-position of the 7-diazonium fluoroborate in acetic acid afforded the 7-fluoro compound (XL; R = F)[44]. Irradiation of 11a-bromo-6-de-methyl-6-deoxytetracycline (XLI; R = Br) in several solvents has been studied[45]. The 7-bromo compound (XL; R = Br) is formed in methanol or acetic acid, whereas in acetonitrile solution dehydrobromination to the 5a,6-anhydro level (XLIII) occurs. Competition experiments with α-naphthol show that the first of these processes is intermolecular. The

10

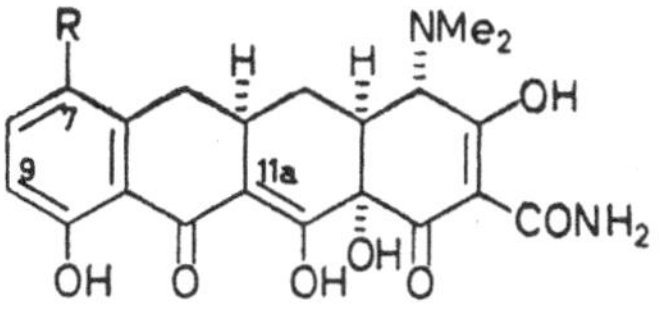

(XL)

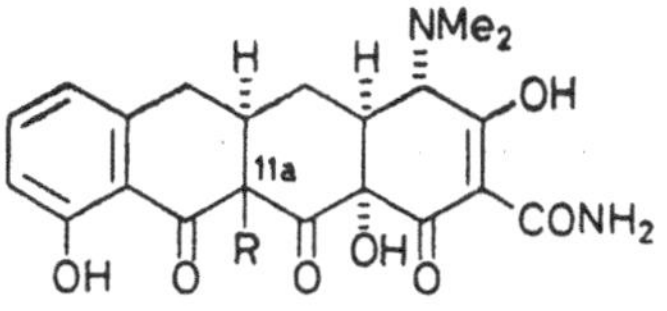

(XLI)

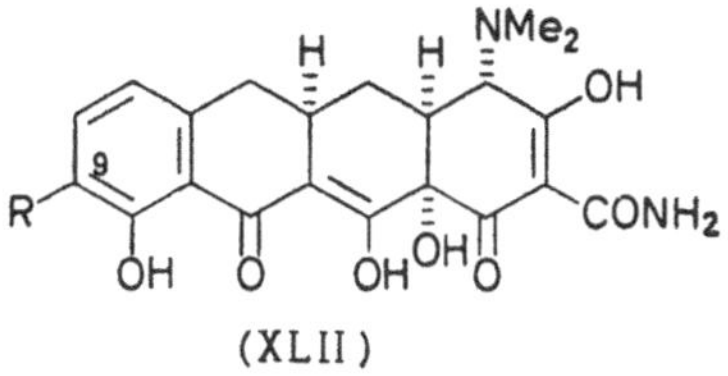

(XLII)

11a-chloro and fluoro compounds (XLI; R = Cl) and (XLI; R = F) are quite stable to all but reducing agents[46]; the latter compound is prepared by the action of perchloryl fluoride on (XXXIX). 2-Nitriles and their 10-benzene sulphonates in the 6-deoxy series can be converted[47] to N-alkyl amides (as XLIV) by the Ritter reaction[48] (isobutene/acetic acid/sulphuric acid). The 6-deoxytetracyclines share the properties of 4-dedimethylamination, 12a-deoxygenation and C_4-epimerization with the parent tetracyclines.

Participation of the 6-hydroxy group was noted during tetracycloxide formation. Under certain reaction conditions, bridging from the 6 to both 11 and 12 carbonyl functions has been observed. For example the action of base on tetracycline causes 11,11a cleavage to isotetracycline (XLV; R = H). Reaction of tetracycline with N-chlorosuccinimide or perchlorylfluoride affords the 11a-chloro or -fluoro-6,12-hemiketals (XLVI) and (XLVII), which are important intermediates for another class of dehydration product, the 6-methylene tetracyclines[49] (see below).

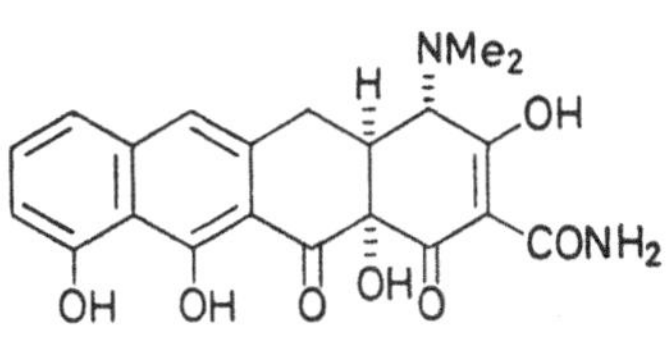

(XLIII)

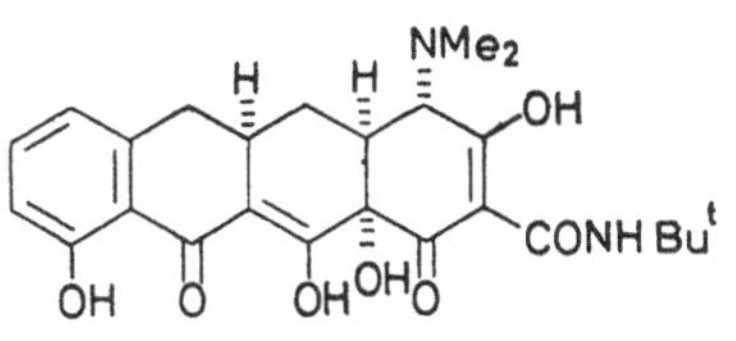

(XLIV)

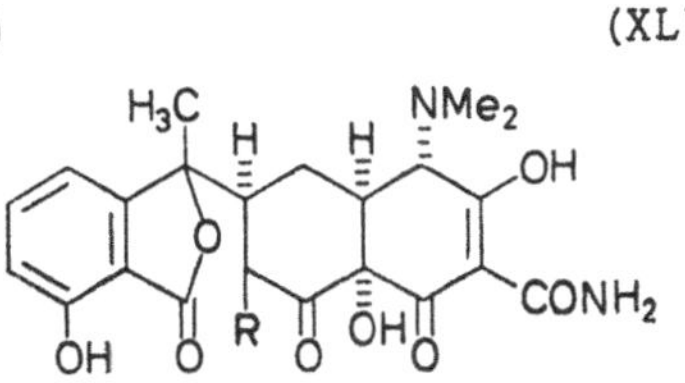

(XLV)

11

The action of basic perchlorylfluoride on 6-deoxy tetracyclines affords simple 11a-fluoro derivatives e.g. (XLVIII) and (XLIX), showing carbonyl absorption in the infrared above 1670 cm^{-1}. On the other hand 6-hydroxytetracyclines (tetracycline, aureomycin) with the same reagent form the 6,12-hemiketals (as XLVII) showing no carbonyl stretching frequency above 1665 cm^{-1}[49]. Indirect proof of the C$_6$-stereochemistry in the 6-demethyl series is provided by the analogous formation of 6-demethyl-6,12-hemiketals[20].

6-*Methylene Tetracyclines*[49]—With methanolic hydrogen chloride 11a-chlorotetracycline-6,12-hemiketal (XLVI) is transformed into iso-chlorotetracycline (XLV; R = Cl) whereas in anhydrous hydrogen fluoride a new class of derivative, the 6(13)-methylenetetracyclines, is produced e.g. (XLVII) → (L; X = H). This exocyclic loss of water is preferred to the well-known 5a,6 (endo) dehydration possibly because hemiketal ring opening is rate controlling, preceding dehydration, and the success of 5a,6 elimination depends on the presence of an 11,11a-double bond to provide driving force for ring C aromatization. In this connection it is noteworthy that 11a-fluoro 6-demethyltetracycline (LI), where *only* 5a,6-dehydration is possible, is stable to HF. However, an 11a-block is not mandatory for the synthesis of methylene compounds as the sulphate ester (LII) can be converted into (LIII)[49]. The

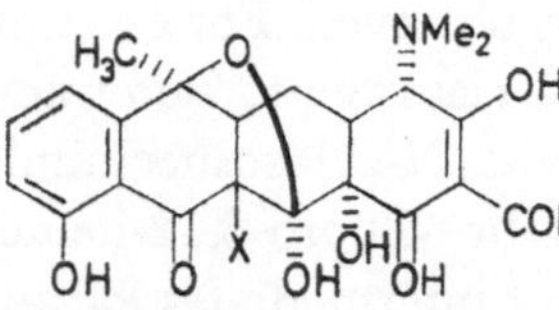

(XLVI) X = Cl
(XLVII) X = F

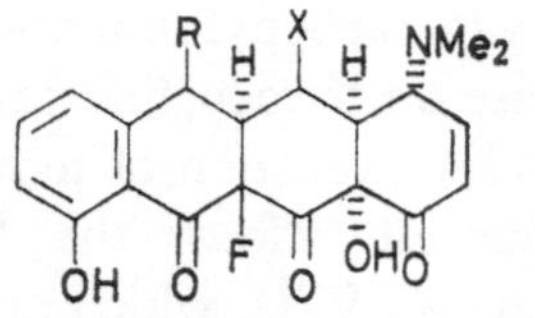

(XLVIII) R = H; X = H
(XLIX) R = CH$_3$; X = OH

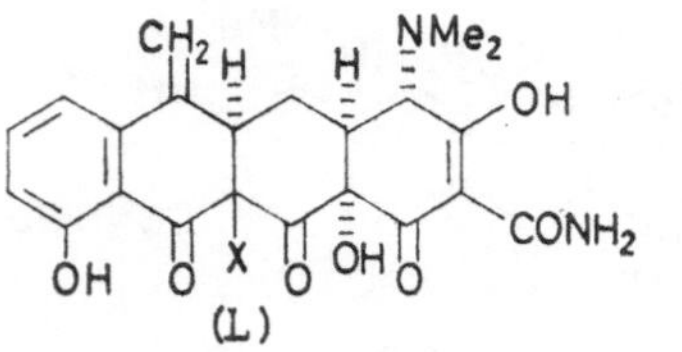

(L)

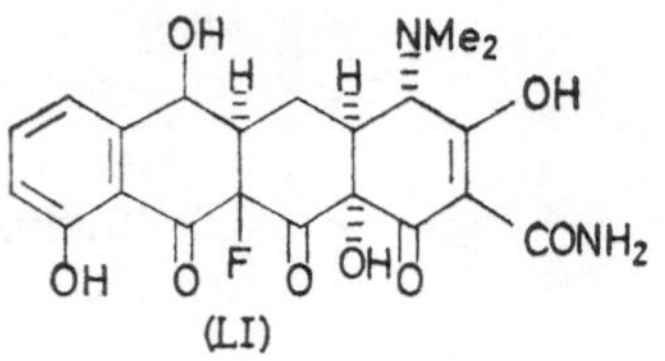

(LI)

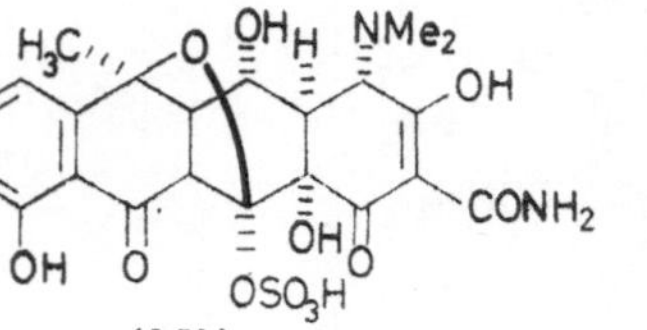

(LII)

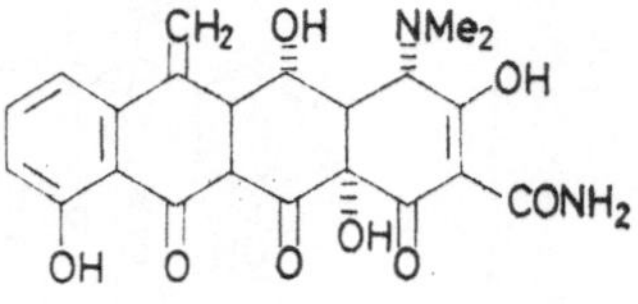

(LIII)

12

11a-halomethylenetetracyclines can be catalytically reduced to methylene tetracyclines (L; X = H). Acid rearrangement to 5a,6-anhydrotetracycline, ozonolysis and catalytic reduction to α (XXXIXb) and β (XXXIXa) -6-deoxytetracycline comprise the main structural evidence for methylene tetracycline (L; X = H) itself[49].

Addition of mercaptans to the exo double bond has been studied intensively. Examples of 13-alkyl, -aryl, -aralkyl and -acyl-α-6-deoxy tetracyclines produced by this route are (LIV) → (LIX). The reaction is typical of free radical mercaptan-olefin addition in that it follows stereospecific anti-Markovnikov orientation. Treatment of the benzyl-sulphoxide (LX) with hydrochloric acid affords (LXII) and (LXIII), the latter possibly via (LXI)[49, 50]. The equatorial or α-configuration for the benzyl mercaptan adduct has been proved by catalytic reduction[50]

	R	Y
(LIV)	Ph	H
(LV)	Ph	OH
(LVI)	PhCH$_2$	H
(LVII)	PhCH$_2$	OH
(LVIII)	CH$_3$CO	H
(LIX)	CH$_3$CO	OH

to α-6-deoxytetracycline[22] (XXXIXb). In the catalytic reduction of 6-(13)-methylene-5-hydroxytetracycline a 1:1 mixture epimeric at C_6 is formed (cf. XXXIXa, b) whilst hydrogenation of 11a-fluoro-6-methylene-5-hydroxytetracycline (LXIV) gives predominantly the β-epimer (LXV). Models clearly show that the curvature of the molecule (LXIV) is such that attack from the α-face is favoured at C_6[50].

Oxidation at C_6—Anticipating the biological conversion of 5a,6-anhydrotetracyclines proper, it was shown[51] that 7-chloro-5a,6-anhydrotetracycline (LXVI) undergoes a smooth photosensitized oxidation with molecular oxygen most probably via (LXVII) to give good yields of 6-deoxy-6-hydroperoxy-5,5a-dehydrotetracycline (LXVIII). This hydroperoxide is easily reducible to 7-chloro-(5a,11a)-5,5a-dehydrotetracycline (LXIX) which has been further reduced to tetracycline[27] (p. 21). The high yields and stereospecifity at C_6 augur well for synthetic studies centred on the anhydrotetracyclines. Other

13

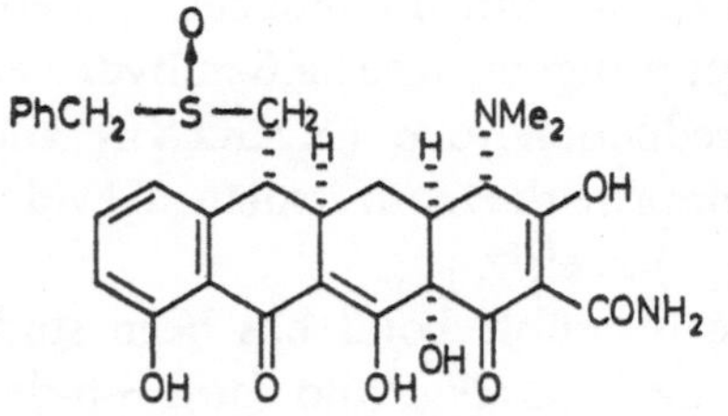

(LX) 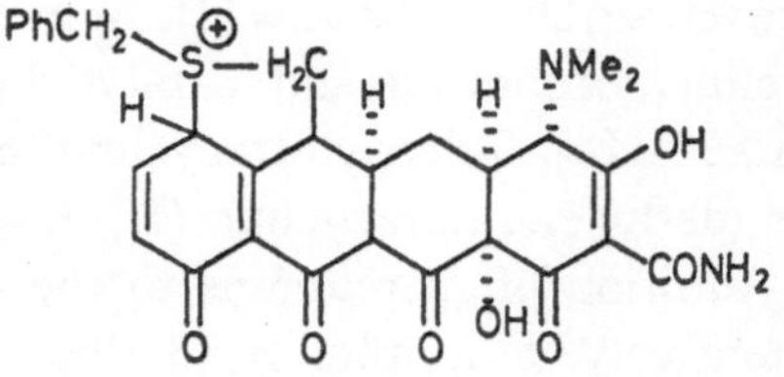 (LXI)

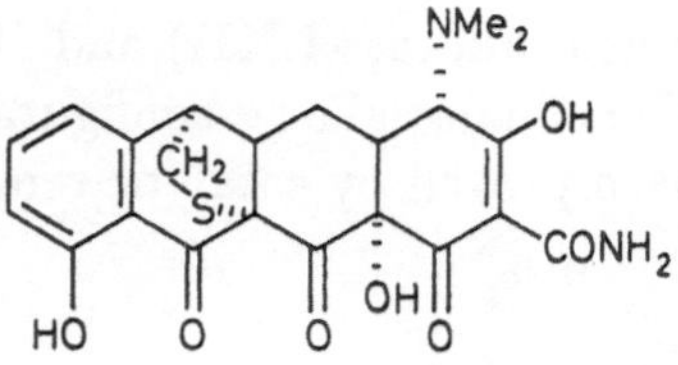

(LXII) 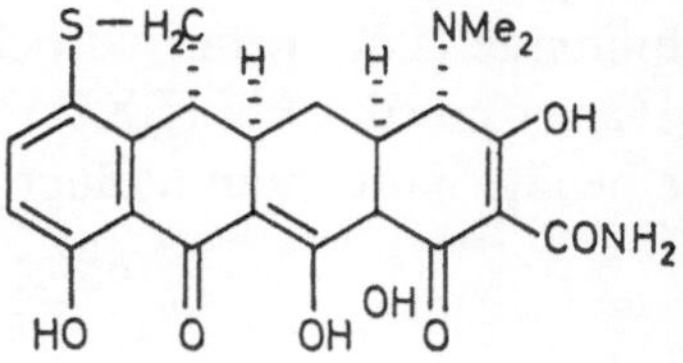 (LXIII)

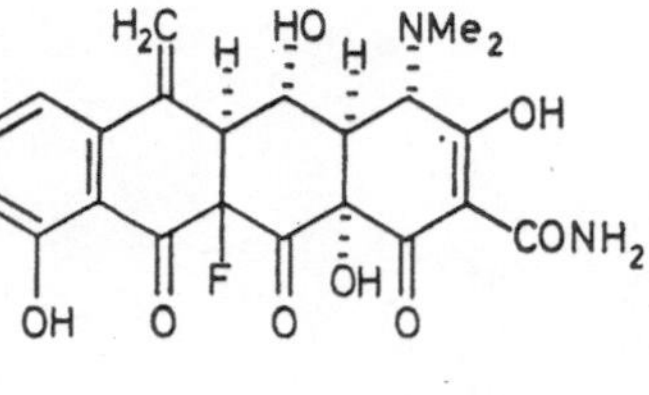

(LXIV) 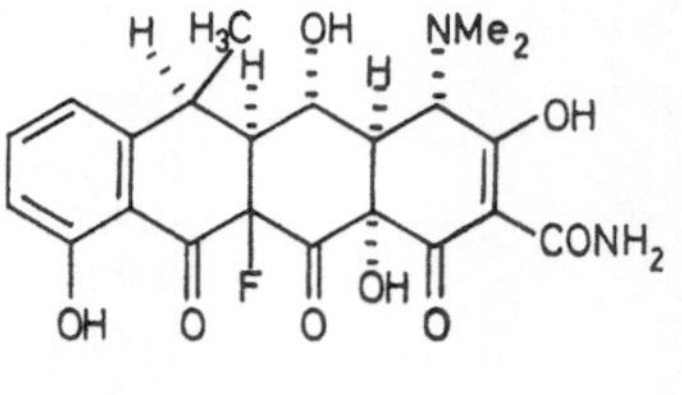 (LXV)

anhydrotetracyclines which undergo the photo-addition of oxygen are the 7-chloro-N-t-butyl[52], 7-chloro-N-di-t-butyl[52] and 7-chloro-4-dedimethylamino[52] derivatives.

Anhydrotetracycline oxidizes at C_6 in much poorer yield, which may be explained by the solubility of the resultant 6-hydroperoxide. Further reduction of the latter has given tetracycline, identified by chromatography[52] and by isolation[53] (1–10 per cent yields). It has also been found that the 6-position of 7-chloroanhydrotetracycline is attacked by lead tetraacetate[53a] (WESSELY oxidation) monopersulphuric acid[53a] (BAMBERGER oxidation) and, under certain conditions by N-bromo- and N-chloro-succinimides[53b].

C_7,C_9—Reactions such as electrophilic aromatic substitution which cause degradation at the tetracycline level, are confined to the 6-deoxy series (see above). The halogen group may be hydrogenolysed from the

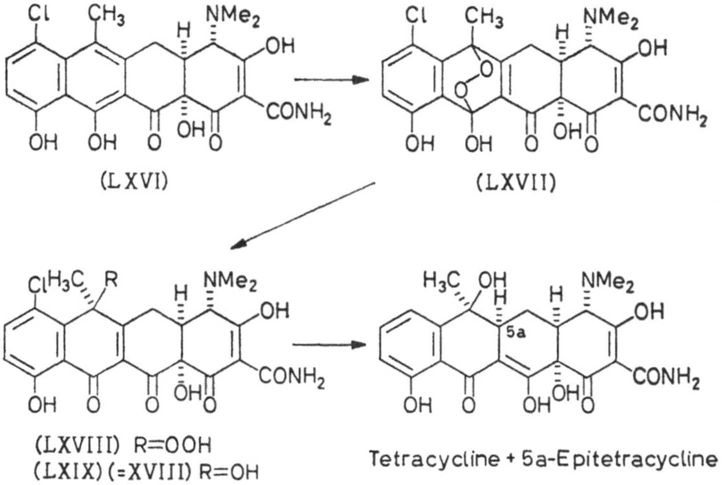

(LXVI)   (LXVII)

(LXVIII) R=OOH
(LXIX) (=XVIII) R=OH Tetracycline + 5a-Epitetracycline

7-halo-tetracyclines[27]. Reintroduction of chlorine to the 7-position has been observed for example in the case of anhydrotetracycline, using either chlorine in acetic acid or sulphuryl chloride[53b], whilst bromination at the 9-position of dedimethylaminoanhydrotetracycline has been reported[54].

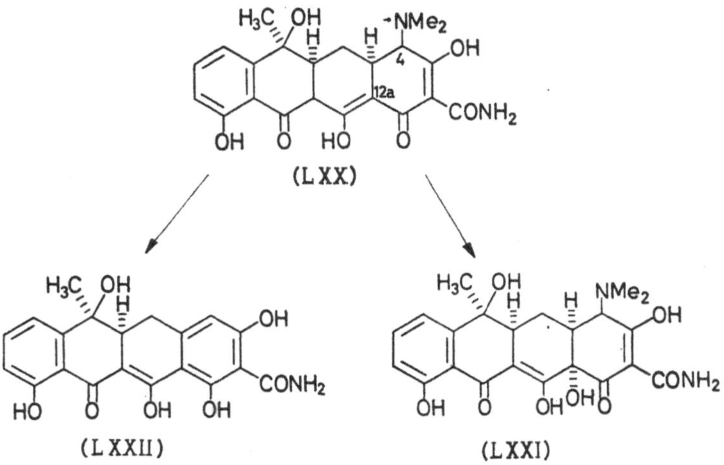

(LXX)

(LXXII)   (LXXI)

C_{11a}—Introduction of halogen at C_{11a} was discussed above in connection with the reactions of perchlorylfluoride and of N-chlorsuccinimide with various tetracyclines. Competition experiments with 12a-deoxydedimethylaminotetracycline show that bromination[54] takes place in the order 12a > 11a.

15

C_{12a}—During the original structural studies[12] it was observed that zinc-ammonia treatment of tetracyclines reductively removed the 12a-hydroxyl function. The reaction occurs with epimerization at C_4[55, 56]. Stereospecific reintroduction of the C_{12a}-hydroxyl group, a step simulating part of the biosynthesis sequence (see p. 24), has been achieved. Thus, 12a-deoxy-4-epi-tetracycline (LXX) on catalytic hydroxylation gives 4-epi-tetracycline (LXXI) with platinum and oxygen in dimethylformamide solution[56]. With perbenzoic acid 20 per cent of the product was dedimethylamino-4a,12a-dehydrotetracyc-line[55] (LXXII). 12a-Hydroxylation also takes place when the oxidant is sodium nitrite in air[59] and in some cases this reagent also produces some 11a-hydroxy compound.

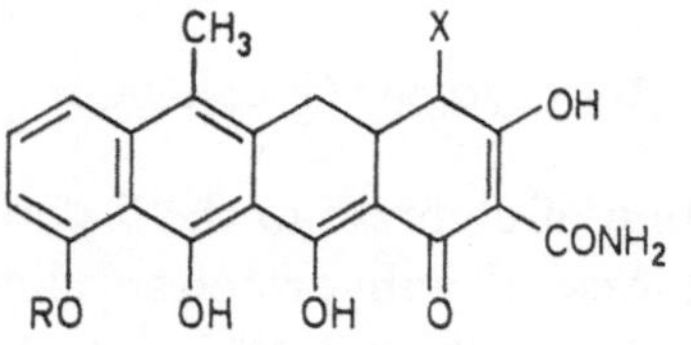

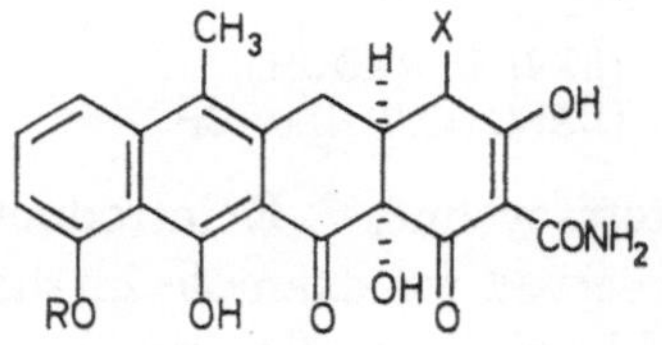

(LXXIII)   R=X=H

(LXXV)   R=CH₃;X=H

(LXXIX)   R=H; X=NMe₂

(LXXIV)   R=X=H

(LXXVI)   R=CH₃; X=H

(LXXX)   R=H; X=NMe₂

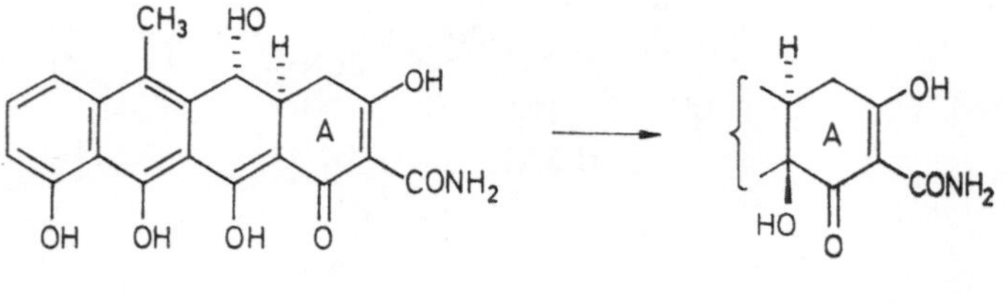

(LXXVII)

(LXXVIII)

Several examples of 12a-hydroxylation of 12a-deoxyanhydrotetra-cyclines e.g. (LXXIII) → (LXXIV) and (LXXV) → (LXXVI) have been recorded[56, 59]. Reagents include perbenzoic acid[58] and platinum–oxygen[56]. With dedimethylamino-12a-deoxyanhydro-5-hydroxytetra-cycline (LXXVII), chloroperbenzoic acid furnishes dedimethylamino-12a-*epi*-anhydro-5-hydroxytetracycline (LXXVIII)[20].

12a-Deoxyanhydrotetracycline (LXXIX) a compound of interest in tetracycline biosynthesis (p. 23) has been rehydroxylated stereo-specifically at C_{12a} to anhydrotetracycline (LXXX) using platinum–oxygen in benzene[53b]. During many of these oxidative experiments

N-oxide and tetracycloxide formation probably contribute to the general difficulty of achieving good reaction conditions.

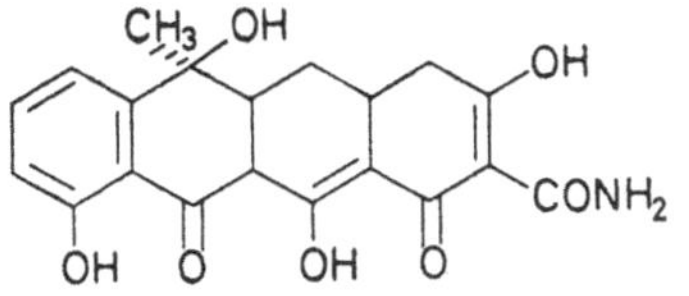

(LXXXI)

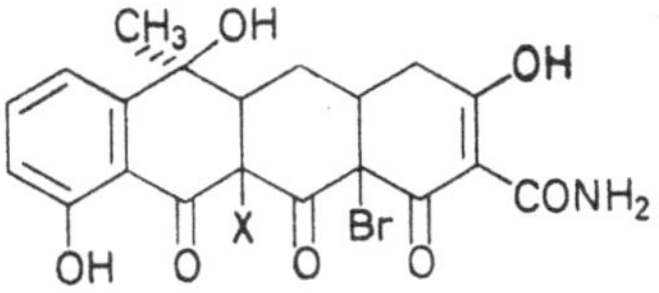

(LXXXII) X = Br
(LXXXIII) X = H

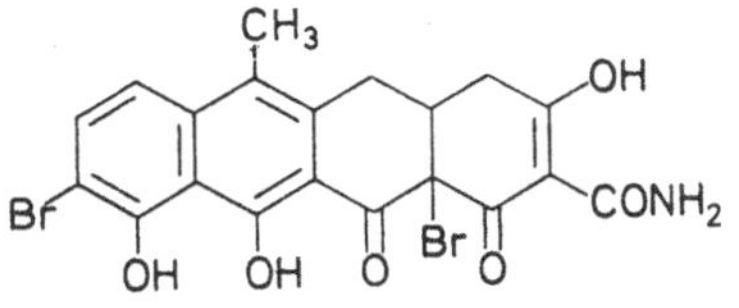

(LXXXIV)

Substitution by halogen at the activated 12a-position in 12a-*desoxy* compounds has also been studied. With two equivalents of N-bromo-succinimide, 4-dedimethylamino-12a-*desoxy*tetracycline (LXXXI) forms the 11a,12a-dibromide (LXXXII) and with one equivalent the 12a-bromo compound (LXXXIII). Treatment of the dibromide (LXXXII) with HBr affords the 9-bromo-anhydrotetracycline (LXXXIV)[54]. Base-catalysed elimination of hydrogen bromide from the 12a-monobromide (LXXXIII) gives the ring-A aromatic 4a,12a-dehydrotetracycline[54] (LXXXVII; R=H). The latter compound is also obtained by pyrolysis (*cis*-4a,12a-elimination) of dedimethylamino 12a-O-formyltetracycline[57] (LXXXV). The O-formyl compounds as (LXXXVI) are best prepared by treatment of a 12a-hydroxytetracycline using a formic-acetic acid mixture[57] and may be catalytically hydrogenated to the corresponding 12a-deoxy compounds, or pyrolysed to 4,4a-dehydro compounds [as (LXXXVII)].

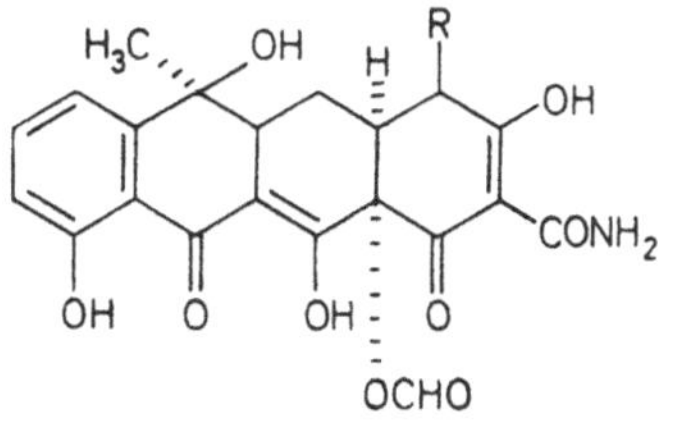

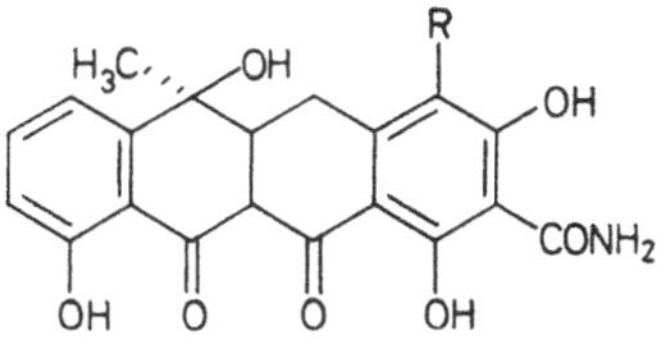

(LXXXV) R = H

(LXXXVI) R = NMe₂

(LXXXVII)

17

BIOSYNTHESIS OF TETRACYCLINES

The biosynthetic pathways to the tetracycline antibiotics have been the subject of considerable speculation and experimental scrutiny since the structural elucidation of the original members in 1952–54.

Structural analysis revealed the typical oxygenation pattern of an acetate-derived phenolic compound[61, 62] and the non-acetyl derivative, 1,3,10,11,12-pentahydroxynaphthacene (LXXXVIII) was suggested as an intermediate in the biosynthetic pathway[62]. Tracer studies[63], using 2-^{14}C-acetic acid, led to the suggestion that the ring structure of 5-hydroxytetracycline (V) was largely, but not entirely, built up from acetate units, with glutamate presumably supplying carbon atoms 2,3,4,4a and the carboxamide C atom (broken lines, *Figure 1*). It was

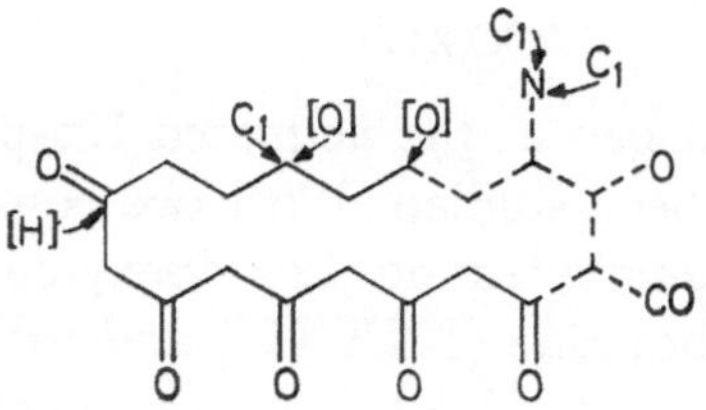

Figure 1. Biosynthetic scheme for tetracycline

also concluded that the methyl groups of the dimethylamino function and the C_6 methyl were derived from methionine thus removing the possibility of a mixed acetate-propionate pathway.

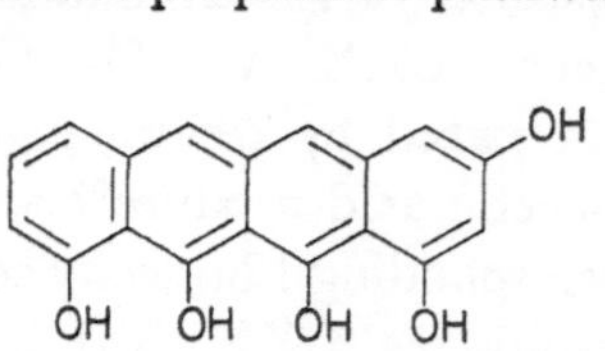

(LXXXVIII)

Later incorporation studies[64] demonstrated that the total tetracyclic nucleus of 5-hydroxytetracycline was derived from acetate and that radioactive glutamate was implicated in some way. In addition, feeding experiments with ^{14}C-bicarbonate demonstrated that the total radioactivity in the isolated 5-hydroxytetracycline was localized in the carboxamide group—the latter was therefore introduced via a carboxylation reaction[64]. Further experiments with carboxyl labelled malonate indicated that, in keeping with the biogenesis of fatty acids and certain

18

phenolic compounds, malonate was the true condensing unit, and importantly that the carboxamide group contained 10–20 per cent of the total radioactivity. It was suggested[64] therefore that the tetracycline nucleus was built up by the condensation of malonate units initiated by malonamyl coenzyme A (see *Figure 2*). The use of the malonate as a

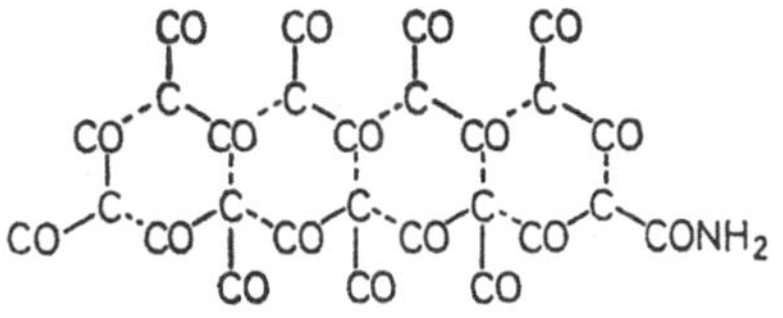

Figure 2. *Malonate units in tetracycline biosynthesis*

starter unit in the construction of polyketide chains, although unusual, has been demonstrated in the case of cycloheximide[65]. Also, the co-occurrence of 2-acetyl-2-decarboxamido-oxy-tetracycline[25] in *S. aureofaciens* and of similar compounds (XIII) and (XV) with chlorotetracycline in a mutant of the same microorganism[26], tends to support this

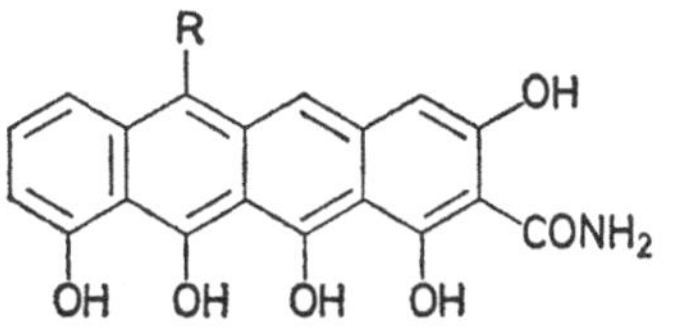

(LXXXIX) R = CH₃
(XC) R = H

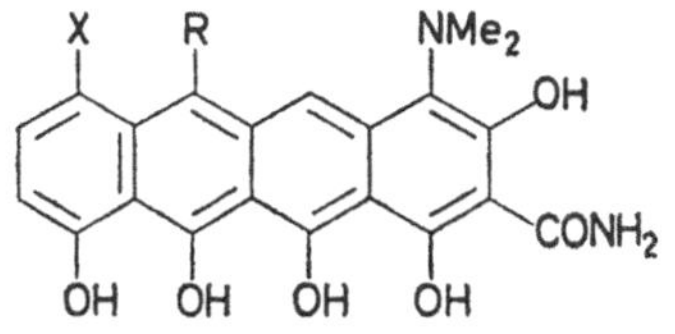

(XCI) X = H; R = CH₃
(XCII) X = Cl; R = H

proposal. The nucleus of the acetyl tetracycline could obviously be initiated by an acetate unit, followed by consecutive condensation of malonate units.

These results and the earlier suggestion that the naphthacene (LXXXVIII) could function as an intermediate in the biosynthetic process, led McCORMICK[66, 67] to study suitably substituted naphthacenes and tetracycline derivatives for precursor activity. In this way he was able to confirm the suggested intermediacy of fully aromatic compounds and specifically to demonstrate that methylpretetramid (LXXXIX) and pretetramid (XC) were normal intermediates in the biosynthesis of the 6-methylated and 6-non-methylated tetracyclines respectively. In addition it was shown that terrarubein (XCI) or 7-chloro-6-*des*methylterrarubein (XCII) did not exhibit significant precursor activity. Later experiments (see below) seem to indicate that N-methylation occurs later in the biosynthetic sequence so that it is still not yet certain whether the corresponding amino derivatives (e.g. XCI; $NMe_2 = NH_2$) could function as normal intermediates. As a result of this study McCormick has also suggested that the 6-methyl group is introduced before, and the 7-chloro substituent after the tetracyclic nucleus is assembled.

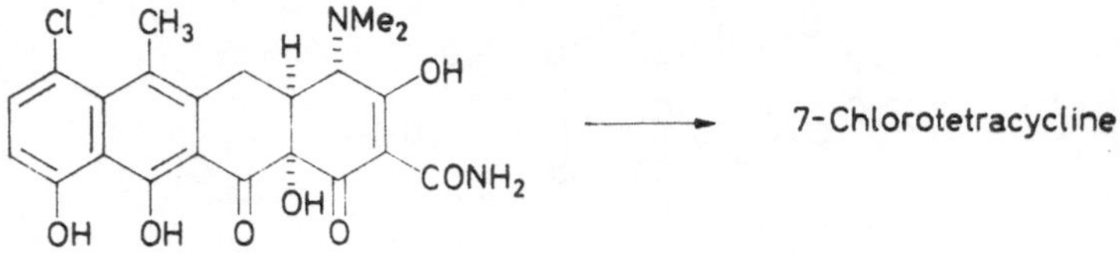

(XCIII) ⟶ 7-Chlorotetracycline

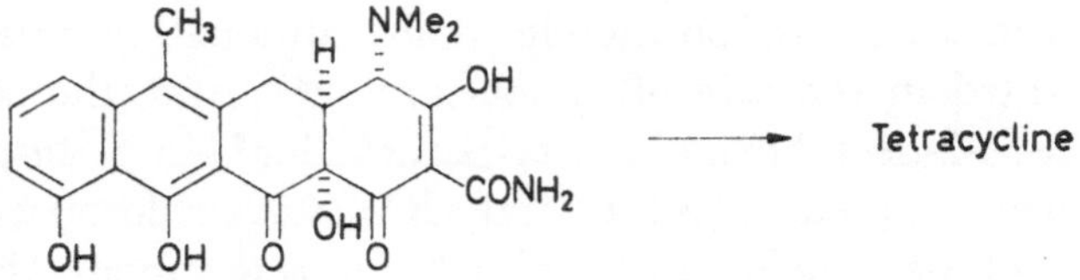

(XCIV) ⟶ Tetracycline

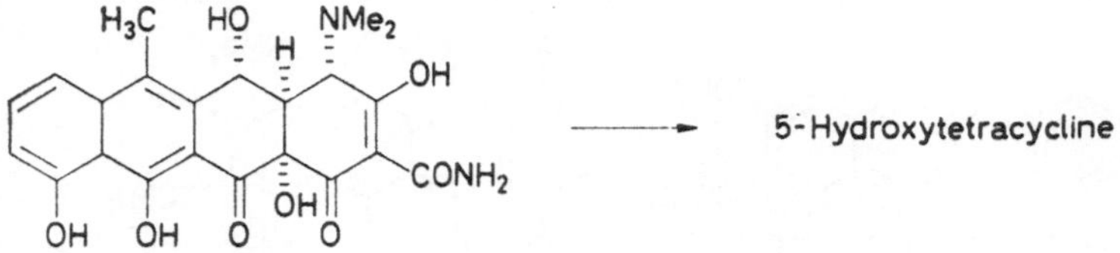

(XCV) ⟶ 5-Hydroxytetracycline

(XCVI) ⟶ 6-Demethyl-7-chloro-tetracycline

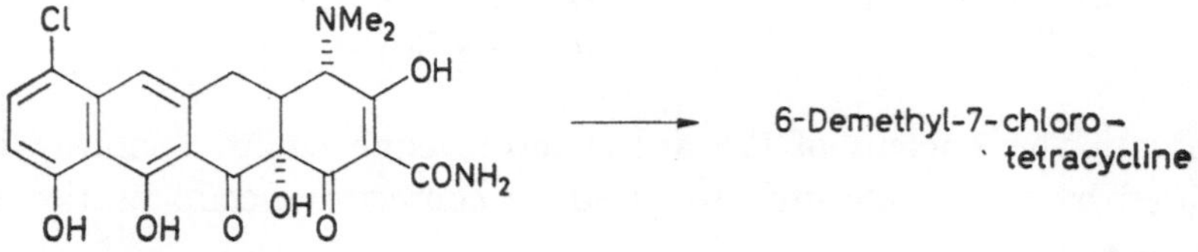

Evidence has also been provided for the final stages in the biosynthesis of the tetracyclines. Thus it has been demonstrated[68] that the 5a,6-anhydro derivatives (XCIII–XCVI) have significant precursor activity and can be converted into the parent tetracyclines as shown by various strains of *S. aureofaciens* and *S. rimosus*, The previous isolation[27] and proved precursor activity[69] of 7-chloro-5a,11a-dehydrotetra-cycline (XVIII) fitted well within this scheme and the terminal stages in the biosynthesis of chlorotetracycline may now be represented as shown (XCIII → XVIII → III). It would appear very probable that, by analogy, the other anhydro precursors are transformed to the parent tetracycline by a similar mechanism. The intervention of (XVIII) in 5-hydroxytetracycline biosynthesis has recently been established[70] (XVIII → XVIIIa → V). It is concluded that direct attack by molecular oxygen is responsible for oxidation at the 5 and 6 positions.

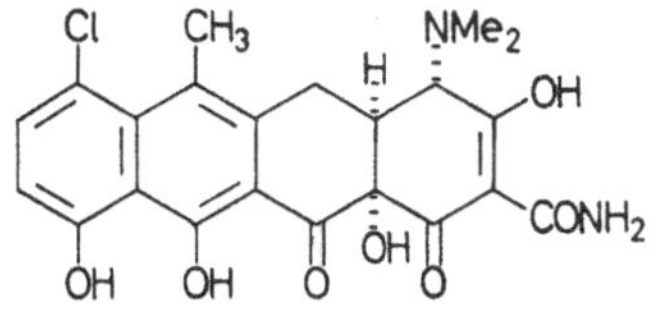

(XCIII)

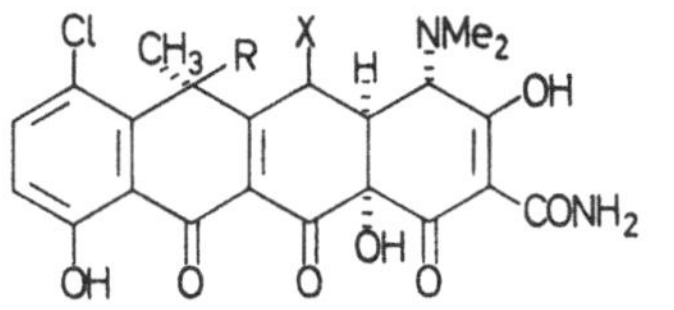

(XVIII) R=OH; X=H
(XCVII) R=OOH; X=H
(XVIIIa) R=OH; X=OH; Cl=H

The reduction of (XVIII) to (III) has been accomplished in the laboratory[27]. *In vitro* studies have also shown that photo-oxygenation of chloranhydrotetracycline (XCIII) produces 7-chloro-5a,11-dehydrotetracycline (XVIII) in good yield via the hydroperoxide (XCVII)[51], a procedure subsequently applied successfully to other anhydro derivatives[52].

The sequence of reactions from methylpretetramid (LXXXIX) and pretetramid (XC) to the anhydrotetracyclines has not been fully elucidated. However, the isolation and demonstrated precursor activity of a series of N-demethylanhydrotetracyclines (e.g. XCVIII) has recently been reported[71] and has provided further insight into the biosynthetic route. The obtention of these important precursors was made possible by growing the mould in media containing antimetabolites of

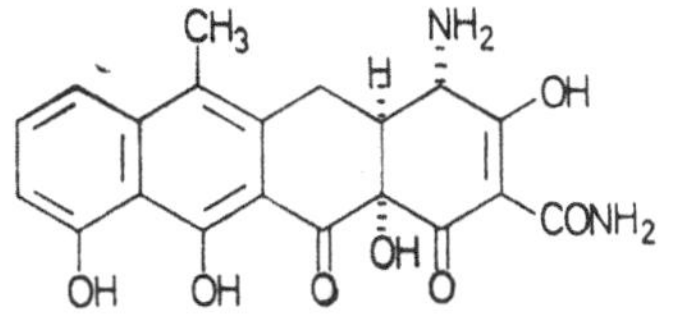

(XCVIII)

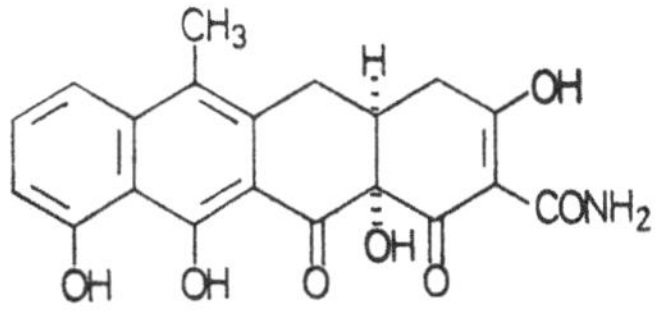

(XCIX)

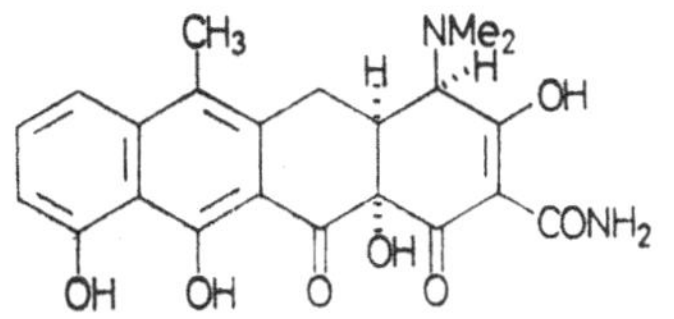

(C)

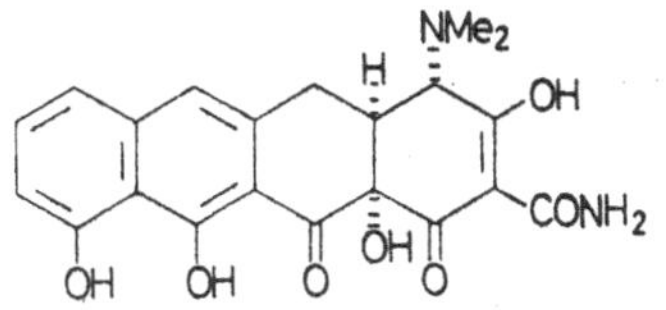

(CI)

21

compounds involved in biological methylation reactions. In the conversion of (XCVIII) into (XCIV) successive methylations occur on the C_4-nitrogen. The 4-epimer of (XCVIII) has been isolated[72] and has been shown not to be a tetracycline precursor. In contrast to C_4-NMe_2 epimerizations, equilibration of (XCVIII) is very slow.

Several interesting conclusions have been drawn from incorporation work using anhydro precursors[68]. The failure of 5a,6-anhydrodedimethylaminotetracycline (XCIX) and 5a,6-anhydro-4-epitetracycline (C) to serve as precursors for tetracycline indicates that the presence of the nitrogen group at C_4 in the proper configuration is essential for hydroxylation at C_6. Also it can be concluded that severe structural requirements are necessary in the conversion of anhydrotetracycline (XCIV) into 5-hydroxytetracycline (V) by *S. rimosus*. For example, 5a,6-anhydro-7-chlorotetracycline (XCIII) and 5a,6-anhydro-6-demethyltetracycline (CI) could not be converted into the corresponding 5-hydroxytetracycline. This is in agreement with the fact that these tetracyclines have not been isolated as metabolic products of this microorganism. It is interesting to note that similar structural limitations are evident in the acetyltetracycline series. 7-Chloro-5-hydroxyacetyltetracycline (CII) and 6-demethyl-5-hydroxyacetyltetracycline (CIII) are not known as metabolites of the microorganism. The evidence provided by these studies indicated that the first step in the conversion of anhydrotetracycline into 5-hydroxytetracycline was hydroxylation at C_5[68]. However, the isolation and proved intermediacy of the new metabolite, dehydrotetracycline (CIV), in 5-hydroxytetracycline biosynthesis now suggests ubiquitous C_6-hydroxylation followed either by reduction or further oxidation at C_5[70].

(CII) X = Cl, R = CH₃
(CIII) X = H, R = H

(CIV)

Although the presence of a 12a-hydroxyl group in the anhydro precursors is not necessary for hydroxylation at C_6 it is nevertheless a fact that the 12a-hydroxyl cannot be introduced at a later stage[68]. Thus 5a,6-anhydro-12a-desoxytetracycline (CV) produces 12a-desoxytetracycline (CVI) only. This would seem to suggest that 12a-hydroxylation precedes the development of the dimethylamino (but not necessarily the amino) function at C_4. The evidence presented so far limits the number of ways in which methylpretetramid can be transformed into N-demethylanhydrotetracycline (XCVIII).

In one route methylpretetramid (LXXXIX) could be transformed to (CVII) via successive hydroxylation at C_4 and C_{12a}. Transamination would then be expected to furnish N-demethylanhydrotetracycline (XCVIII). Alternatively, preliminary amination of the 4-hydroxymethylpretetramid followed by C_{12a} hydroxylation and reduction could

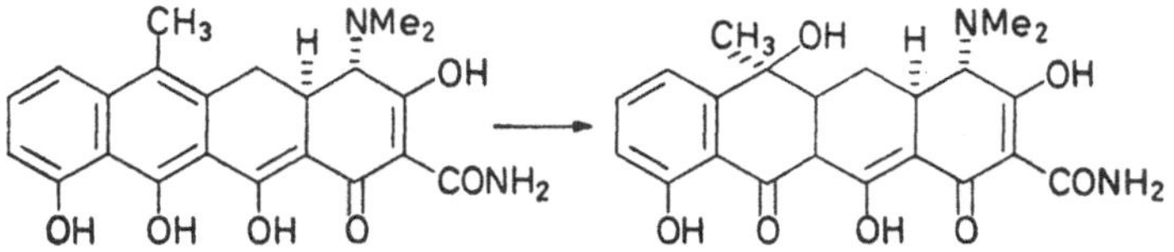

also yield the required anhydro compound (CVIII→CIX→XCVIII). The intermediacy of 4-hydroxymethylpretetramid has been recently confirmed by further studies of McCORMICK[73], who demonstrated that 4-hydroxymethylpretetramid (CIX) is a normal intermediate in the biosynthesis of chlorotetracycline. The laboratory conversions of (LXXXIX) → (CVIII)[34] and of (CV) → (XCIV) → (XCIII) → (XVIII) → tetracycline[39, 53b] have very recently been effected using appropriate sequential combination of the reagents O_2, H_2 and Cl_2 under catalytic control, as part of a biogenetic type synthesis of tetracycline.

The introduction of chlorine into the chlorotetracyclines is an unexceptional process and studies have been conducted on the halide metabolism of *S. aureofaciens* mutants[74]. Other work suggests[66, 68] the introduction of the chloro-substituent after the production of the

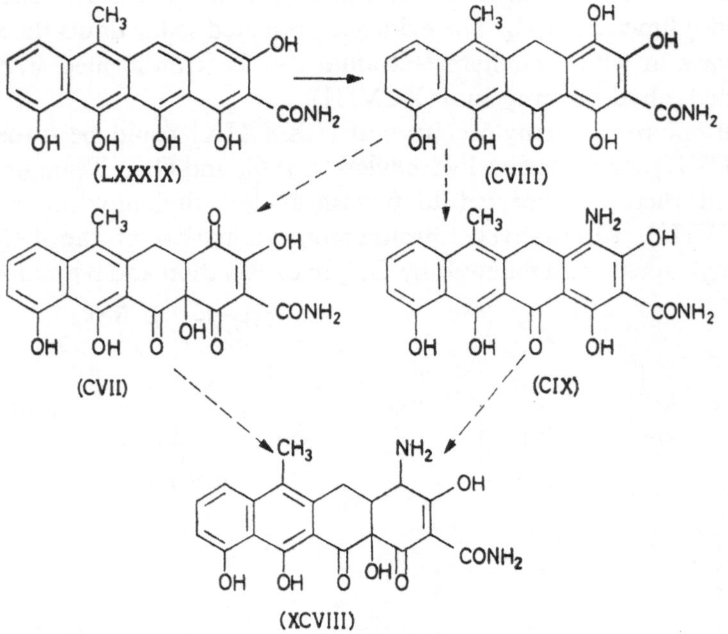

(LXXXIX)

(CVIII)

(CVII)

(CIX)

(XCVIII)

naphthacenic precursor and before the complete elaboration of the 5a,6-anhydrotetracycline framework. Thus a strain which was able to convert chloranhydrotetracycline to chlorotetracycline could only transform anhydrotetracycline to tetracycline. The exact position in the biosynthetic sequence at which chlorine is introduced is still unknown. At the present time the possibilities are rather limited, halogenation occurring between 4-hydroxymethylpretetramid and 7-chlor-N-demethylanhydrotetracycline.

A summary of the biosynthesis of 7-chlorotetracycline is shown in *Figure 3*. An excellent review of the current status of tetracycline biosynthesis has recently appeared[84].

TOTAL SYNTHESIS OF TETRACYCLINES

The completion of the total synthesis of a natural or 'fermentation' tetracycline embodying the elaboration of five* labile centres of asymmetry has not yet been announced. Remarkable progress towards this goal has, however, been made recently on several fronts. Only a description quite out of proportion to the scope of this chapter could properly serve to illustrate the ingenuity and effort behind these

* Or, in the case of terramycin, six.

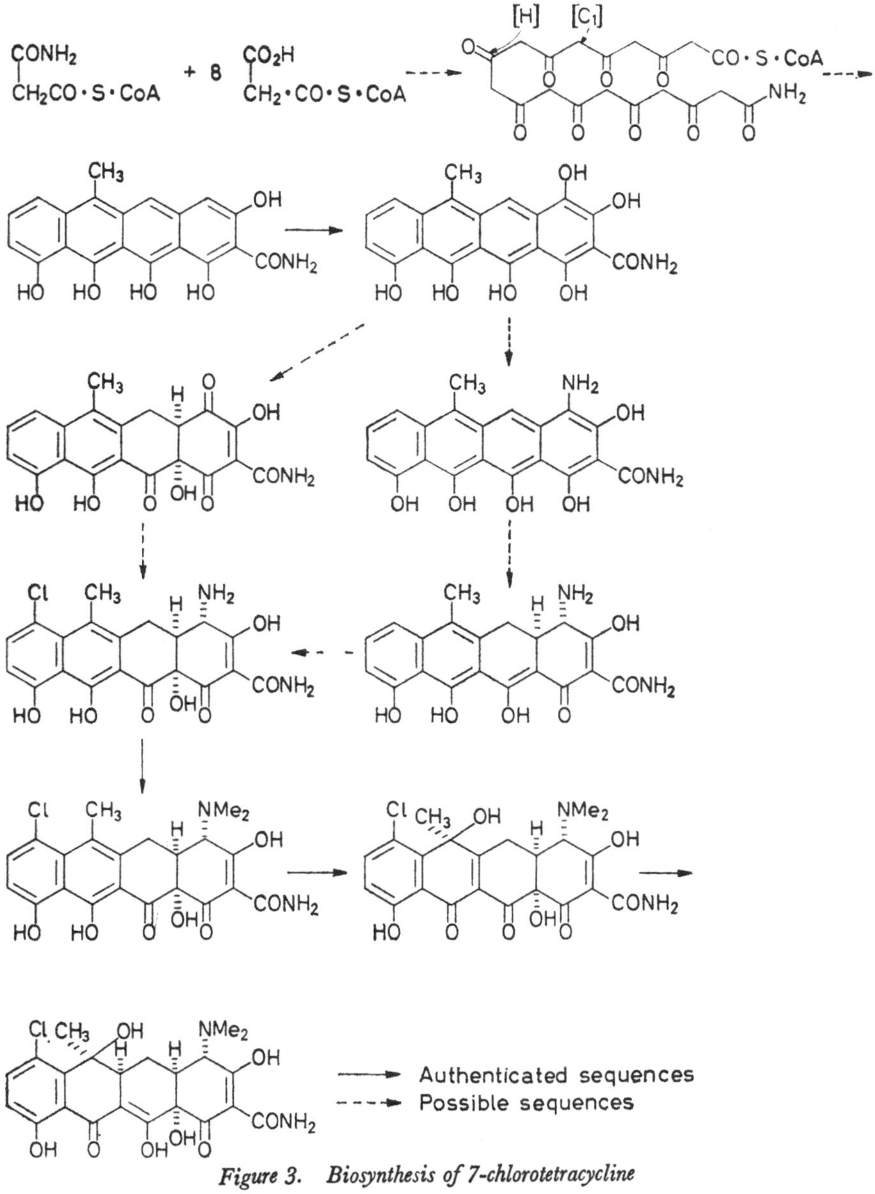

Figure 3. *Biosynthesis of 7-chlorotetracycline*

synthetic studies and we have therefore chosen to present perhaps the most salient features in the summarized charts which follow. Recent reviews have dealt at length with the Wisconsin[8, 9] and Pfizer–Harvard syntheses[76] whilst a more general discussion of the Lederle and Moscow approaches, together with other syntheses of model compounds has been provided in an exhaustive review[6].

3

25

The Braunschweig–Madison Syntheses

From the notable progress along several fronts of tetracycline synthesis made by H. Muxfeldt and his colleagues (now at the University of Wisconsin) we have chosen the syntheses of (*a*) (±)-dedimethylamino 5a,6-anhydro-7-chlorotetracycline and (*b*) a remarkable and elegant preparation of (±)-6-deoxy-6-demethyltetracycline, the fully bio logically-active degradation product of tetracycline via azlactone formation.

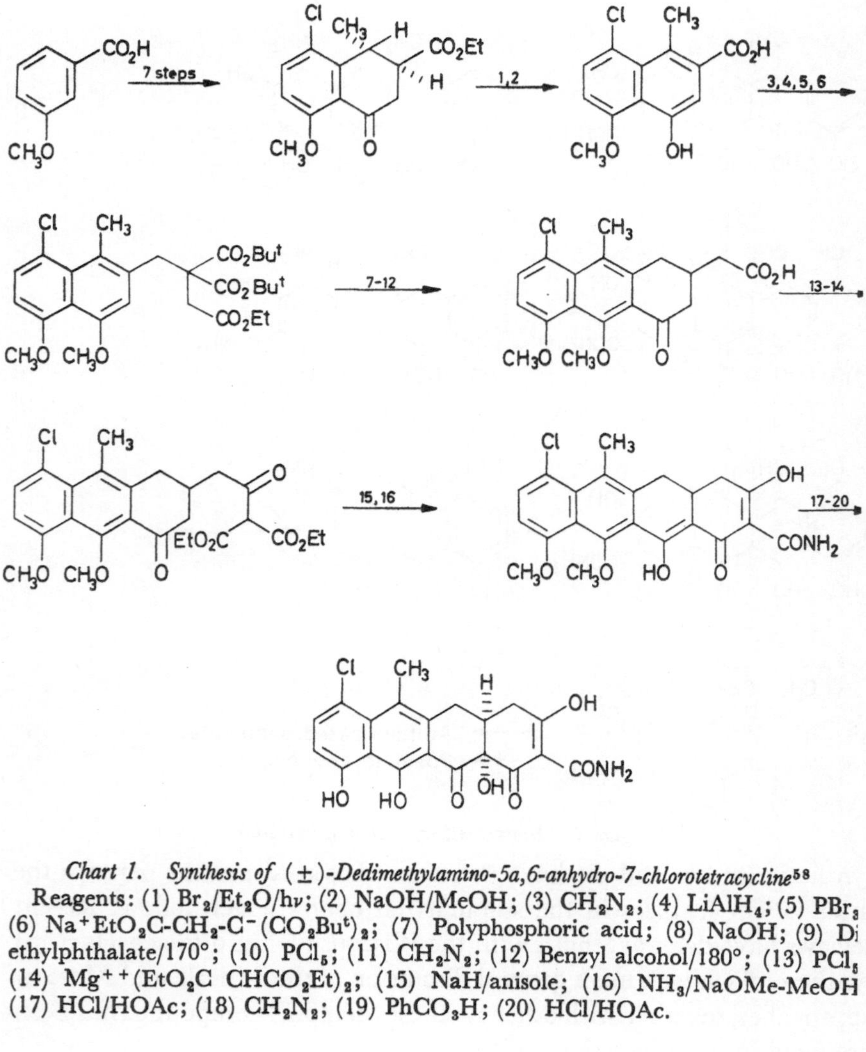

Chart 1. Synthesis of (±)-Dedimethylamino-5a,6-anhydro-7-chlorotetracycline[58]
Reagents: (1) $Br_2/Et_2O/h\nu$; (2) NaOH/MeOH; (3) CH_2N_2; (4) $LiAlH_4$; (5) PBr_3
(6) $Na^+EtO_2C\text{-}CH_2\text{-}C^-(CO_2Bu^t)_2$; (7) Polyphosphoric acid; (8) NaOH; (9) Di ethylphthalate/170°; (10) PCl_5; (11) CH_2N_2; (12) Benzyl alcohol/180°; (13) PCl_5
(14) $Mg^{++}(EtO_2C\ CHCO_2Et)_2$; (15) NaH/anisole; (16) $NH_3/NaOMe\text{-}MeOH$
(17) HCl/HOAc; (18) CH_2N_2; (19) $PhCO_3H$; (20) HCl/HOAc.

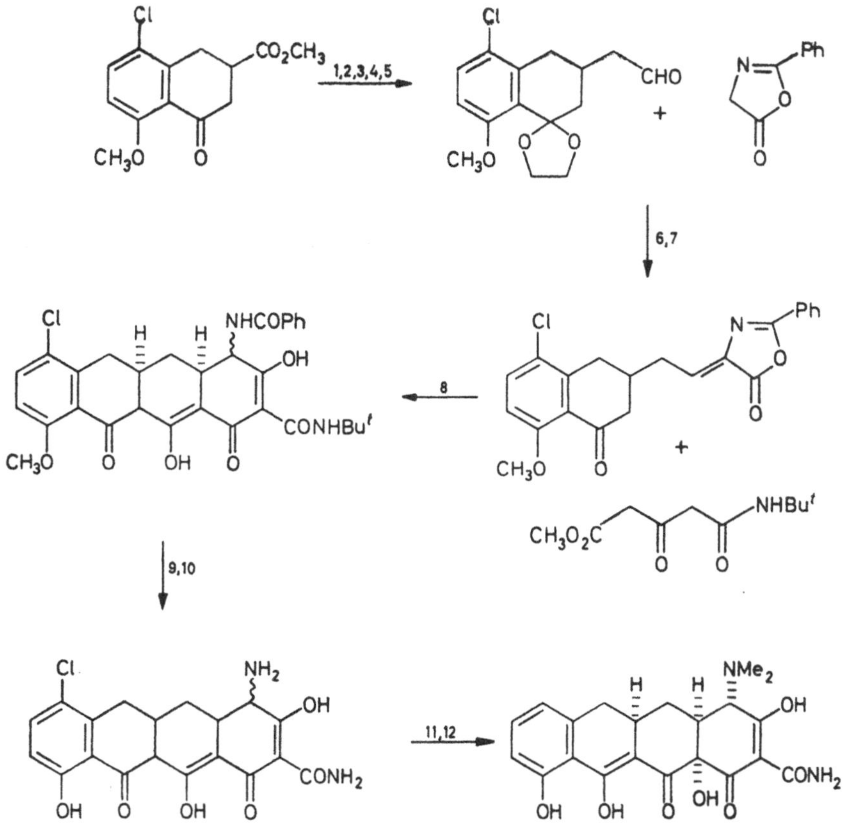

Chart 2. Synthesis of ($\pm$)-6-Deoxy-6-demethyltetracyclines[7b]

Reagents: (1) Ketalization; (2) LiAlH$_4$; (3) MES Cl/Pyridine; (4) KCN/dimethylformamide; (5) LiAl(OEt$_3$)H; (6) Ac$_2$O/Pb(OAc)$_2$; (7) HCl/tetrahydrofuran; (8) NaH/tetrahydrofuran → 2 tautomeric epimers—one C$_4$-epimer separated; (9) Meerwein hydrolysis; (10) HBr/HOAc; (11) CH$_2$O/Pd-C/H$_2$/Et$_3$N; (12) Pt/O$_2$

The Pfizer–Harvard Synthesis

The first total synthesis of the biologically active 6-deoxy-6-demethyltetracycline was announced by the combination of chemists at Groton (L. H. Conover, K. Butler, J. D. Johnston and J. J. Korst), and Harvard (R. B. Woodward). Of particular interest among the varied reactions employed are the successive operations at carefully prepared sites of methylene activity as well as the controlled introduction of the dimethylamino group by preferential β-addition prior to construction and closure of ring A.

27

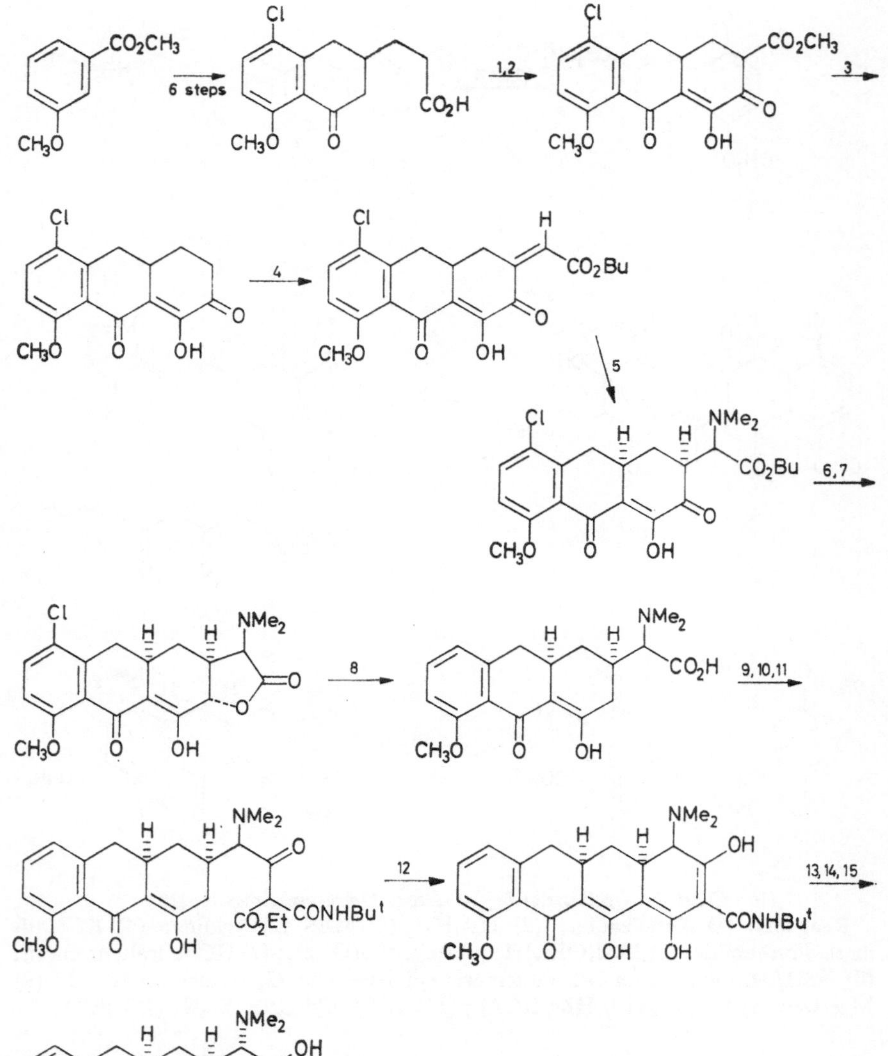

Chart 3. Synthesis of ± 6-Deoxy-6-demethyltetracycline[76]

Reagents: (1) Esterify; (2) $MeO_2C \cdot CO_2Me$/MeOH/NaH/dimethylformamide; (3) HCl/HOAc; (4) $Mg(OMe)_2$/n-butylglyoxalate/toluene; (5) $Me_2NH/-10°$; (6) $NaBH_4$/diglyme; (7) p-toluene sulphonic acid/toluene; (8) Zn/HCO_2H; (9) H_2-Pd/EtOH/Et_3N; (10) $ClCO_2Pr^i$; (11) $Mg^{++}(EtO_2C\text{-}C\overline{H}\text{-}CONHBu^t)_2$; (12)NaH/dimethylformamide; (13) HBr; (14) $CeCl_3/O_2$; (15) Ca^{++}/pH 8·5

28

The Lederle Synthesis

An early success in the construction (with stereochemical control) of the tetracyclic framework is illustrated by the work of T. L. Fields, J. H. Boothe, A. S. Kende and S. Kushner on the synthesis of ($\pm$)-de-dimethylamino-6-demethyl-6,12a-dideoxy-7-chlorotetracycline. Condensations of a type reminiscent of polyketide formation (malonate extension) were used to form the rings in the sequence DCBA.

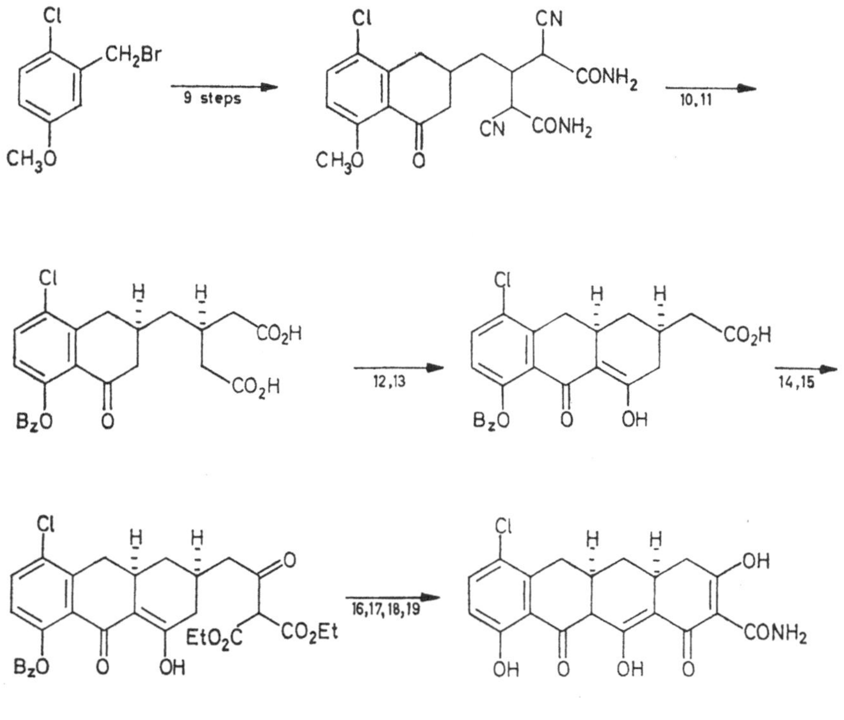

Chart 4. Synthesis of ($\pm$)-Dedimethylamino-6-demethyl-6,12a-dideoxy-7-chlorotetracycline[78]

Reagents: (10) HCl/HOAc; (11) Benzyl chloride/alkali; (12) Ac$_2$O; (13) NaOMe/MeOH; NaH/toluene; (14) ClCO$_2$Et/Et$_3$N; (15) Mg^{++}(EtO$_2$C CH CO$_2$Et)$_2$; (16) NaH/toluene; (17) H$_2$/Pd; (18) Ammonium formate/140°; (19) HCl

The Moscow Synthesis

Several approaches to tetracyclines have been described by M. M. Shemyakin and his co-workers. The route shown in *Chart 5* is of interest for in contrast to the other syntheses[83] a Diels–Alder reaction on naphthaquinone is used to reach the tricyclic series in the first step, allowing a Grignard introduction of the potential 6-methyl-hydroxy system. Another point of note is that 12a-hydroxylation is avoided by retention of an oxygen function at this position from the outset.

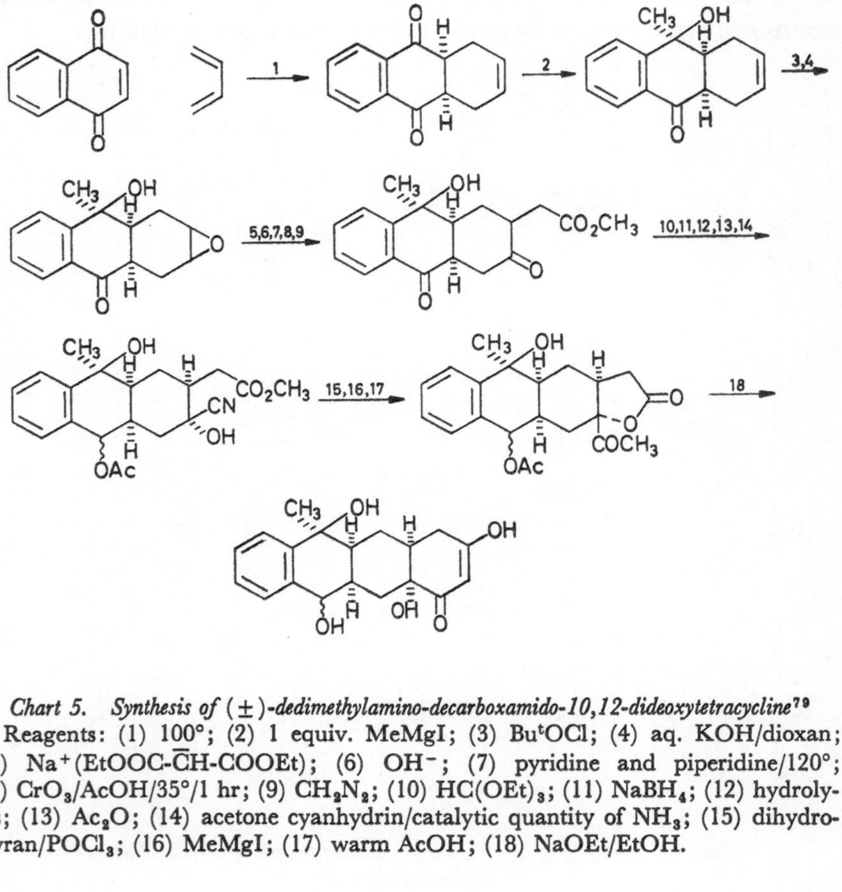

Chart 5.   Synthesis of (±)-dedimethylamino-decarboxamido-10,12-dideoxytetracycline[79]

Reagents: (1) 100°; (2) 1 equiv. MeMgI; (3) Bu^{t}OCl; (4) aq. KOH/dioxan; (5) Na$^+$(EtOOC-$\overline{C}$H-COOEt); (6) OH$^-$; (7) pyridine and piperidine/120°; (8) CrO$_3$/AcOH/35°/1 hr; (9) CH$_2$N$_2$; (10) HC(OEt)$_3$; (11) NaBH$_4$; (12) hydrolysis; (13) Ac$_2$O; (14) acetone cyanhydrin/catalytic quantity of NH$_3$; (15) dihydropyran/POCl$_3$; (16) MeMgI; (17) warm AcOH; (18) NaOEt/EtOH.

CONCLUSION

With the biosynthetic map almost completed and total synthesis reaching its crucial but final stages, it would appear that the main interest in the chemistry of tetracycline antibiotics will centre on the further exploration of chemical reactivity at the various centres, coupled with the production of closely related metabolites (see Chart 6) by Streptomyces mutants or other microorganisms.

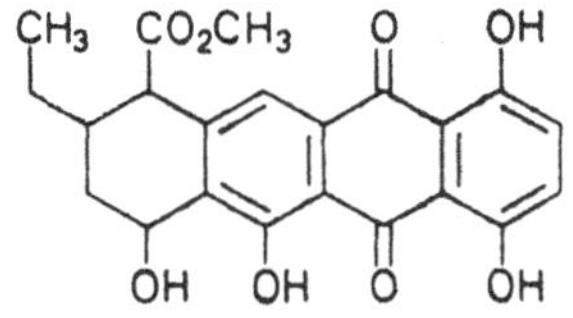

80
ε-Pyrromycinone

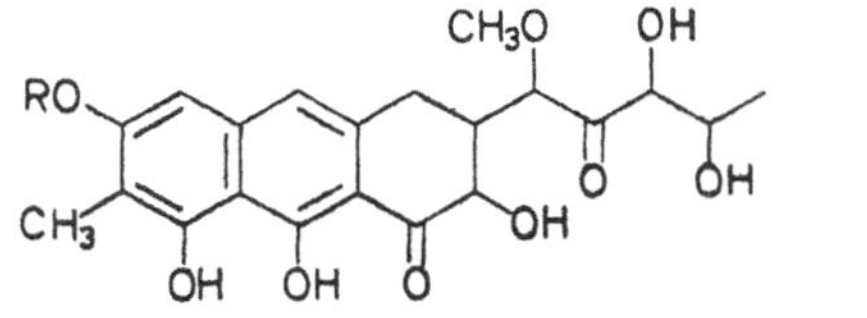

81
Chromomycin A₃ (R=tetrasaccharide)

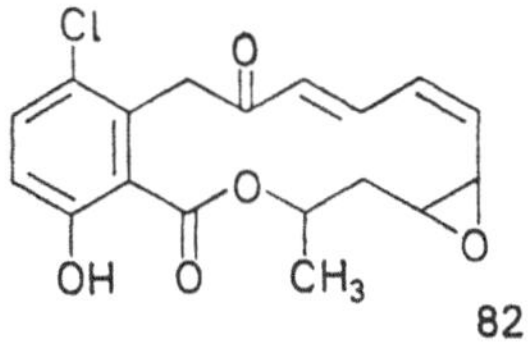

82
Monorden (Radicicol)

Chart 6.   Metabolites related to tetracyclines

REFERENCES

[1] DUGGAR, B. M. *Ann. N.Y. Acad. Sci.* 51 (1948) 177

[2] Tetracycline: DOWLING, H. F. *Medical Encyclopedia Inc.*, New York, 1955. Aureomycin: MUSSELMANN, M. M. *Medical Encyclopedia Inc.*, New York, 1956. Terramycin: LEPPER, M. H. *Medical Encyclopedia Inc.*, New York, 1956

[3] MUXFELDT, H. and BANGERT, R. *Fortschr. Chem. org. Nat. Stoffe.* 21 (1963) 80

[4] VAN TAMELEN, E. E. *Fortschr. Chem. org. Nat. Stoffe.* 16 (1958) 90

[5] REGNA, P. P. In *Antibiotics. Their Chemistry and Non-medical Uses* (ed. Goldberg, H. D.), Van Nostrand, Princeton, 1959, pp. 77–96

[6] BARRETT, G. C. *J. Pharm. Sci.* 52 (1963) 309

[7] BOOTHE, J. H. *Antimicrobial Agents and Chemotherapy—U.S.A.* (1962) 213

[8] MUXFELDT, H. *Angew. Chem.* 74 (1962) 443 (*Int. edn*) 1 (1962) 372

[9] MUXFELDT, H. *Angew. Chem.* 74 (1962) 845

[10] VOGEL, H. *Sci. Pharm.* 24 (1956) 257

[11] PIECHOWSKA, M. *Wladomosci Chemi.* 14 (1960) 779

[12] HOCHSTEIN, F. A., STEPHENS, C. R., CONOVER, L. H., REGNA, P. P., PASTERNACK, R., BRUNINGS, K. J. and WOODWARD, R. B. *J. Amer. chem. Soc.* 74 (1952) 3708

[13] RICHARDS, J. F. and HENDRICKSON, J. B. *Biosynthesis of Steroids, Terpenes, and Acetogenins,* Benjamin, New York, 1964

[14] DOERSCHUK, A. P., BITLER, B. A. and McCORMICK, J. R. D. *J. Amer. chem. Soc.* 77 (1955) 4687

[15] McCORMICK, J. R. D., FOX, S. M., SMITH, L. L., BITLER, B. A., REICHENTHAL, J., ORIGONI, V. E., MULLER, W. H., WINTERBOTTOM, R. and DOERSCHUK, A. P. *J. Amer. chem. Soc.* 78 (1956) 3547

[16] McCORMICK, J. R. D. FOX, S. M., SMITH, L. L., BITLER, B. A., REICHENTHAL, J., ORIGONI, V. E., MULLER, W. H., WINTERBOTTOM, R. and DOERSCHUK, A. P. *ibid.* 79 (1957) 2849

[17] Stephens, C. R., Conover, L. H., Gordon, P. N., Pennington, F. C., Wagner, R. C., Brunings, K. J. and Pilgrim, F. J. *J. Amer. chem. Soc.* 78 (1956) 1515

[18] Donohue, J., Dunitz, J. D., Trueblood, K. N. and Webster, M. S. *J. Amer. chem. Soc.* 85 (1963) 851

[19] Takeuchi, M. and Buerger, M. J. *Proc. nat. Acad. Sci.*, U.S. 46 (1960) 1366

[20] Von Wittenau, M. S., Blackwood, R. K., Conover, L. H., Glauvert, R. H. and Woodward, R. B. *J. Amer. chem. Soc.* 87 (1965) 134

[20a] Dobrynin, V. N., Gurevich, A. I., Karapefyan, M. G., Kolosov, M. N. and Shemyakin, M. M. *Tetrahedron Ltrs* (1962) 901

[21] McCormick, J. R. D., Sjolander, N. O., Hirsch, U., Jensen, E. P. and Doerschuk, A. P. *J. Amer. chem. Soc.* 79 (1957) 4561

[22] Webb, J. S., Broschard, R. W., Cosulich, D. B., Stein, W. J. and Wolf, C. F. *J. Amer. chem. Soc.* 79 (1957) 4563

[23] Boothe, J. H., Green, A., Petisi, J. P., Wilkinson, R. G. and Waller, C. W. *J. Amer. chem. Soc.* 79 (1957) 4564

[24] Doerschuk, A. P., McCormick, J. R. D., Goodman, J. J., Szumski, S. A., Growich, J. A., Miller, P. A., Bitler, B. A., Jensen, E. R., Petty, M. A. and Phelps, A. S. *J. Amer. chem. Soc.* 78 (1956) 1508

[25] Hochstein, F. A., von Wittenau, M. S., Tanner, F. W. and Minai, K. *J. Amer. chem. Soc.* 82 (1960) 5934

[26] Miller, M. W. and Hochstein, F. A. *J. org. Chem.* 27 (1962) 2525

[27] McCormick, J. R. D., Miller, P. A., Growich, J. A., Sjolander, N. O. and Doerschuk, A. P. *J. Amer. chem. Soc.* 80 (1958) 5572

[28] von Wittenau, M. S., Hochstein, F. A. and Stephens, C. R. *J. org. Chem.* 28 (1963) 2454

[29] Dulaney, E. L., Putter, I., Dreschev, D., Chaiet, L., Miller, W. J., Wolf, F. J. and Hendlin, D. *Biochim. biophys. Acta* 60 (1962) 447

[30] Miller, P. A., Satumelli, A., Martin, J. H., Mitscher, L. A. and Bohonos, J. *Biochem. Biophys. Res. Comm.* 16 (1964) 285

[31] Gottstein, W. J., Minor, W. F. and Cheney, L. C. *J. Amer. chem. Soc.* 81 (1959) 1198

[32] Siedel, W., Soeder, F. and Lindner, F. *Munchner Med. Wochenschr.* 17 (1958) 661

[33] Stephens, C. R. *Abstr. Pap. 109th Meet. Amer. chem. Soc.* Dallas, 1956, p. 18M

[34] Noseworthy, M. M. *U.S. Pat.* 3,009,956

[35] Conover, L. H. *Chem. Soc. (Lond.) Spec. Publ.* 5 (1956) 73

[36] Esse, R. C. and Sieger, G. M. *South African Pat.* 63/4791 (1964); Esse, R. C., Lowery, J. A., Tamoria, C. R. and Sieger, G. M. *J. Amer. chem. Soc.* 86 (1964) 3874, 3875

[37] Blackwood, R. K. and Stephens, C. R. *J. Amer. chem. Soc.* 86 (1964) 2736

[38] cf. Leonard, N. J. and Hay, A. S. *J. Amer. chem. Soc.* 78 (1956) 1984; Fuller, R. *Chem. Rev.* 63 (1963) 21; Ellis, A. J. and Soper, F. G. *J. chem. Soc.* (1954) 1750

[39] Scott, A. I., Money, T., Neilson, T., Ramanathan, J. D. and Yalpani, M. Reported in part at the *Amer. chem. Soc. Medicinal Chemistry Symposium, Minneapolis*, 1964 and in *Karl Folkers Lecture Series*, Madison, 1964; to be published

[40] Kolosov, M. N., Shemyakin, M. M., Khoklov, A. S. and Berlin, Ya. A.

In *Chemistry of Antibiotics*, U.S.S.R. Acad. Sci., Moscow, Vol. I, pp. 180–268

[41] MUXFELDT, H. *Angew Chem.* 74 (1962) 443; MUXFELDT, H., ROGALSKI, W. and STRIEGLER, K. *Chem. Ber.* 95 (1962) 2581

[42] PETISI, J., SPENCER, J. L., HLAVKA, J. J. and BOOTHE, J. H. *J. Med. Pharm. Chem.* 5 (1962) 538

[43] HLAVKA, J. J., SCHNELLER, A., KRAZINSKI, H. and BOOTHE, J. H. *J. Amer. chem. Soc.* 84 (1962) 1426

[44] HLAVKA, J. J., KRAZINSKI, H. and BOOTHE, J. H. *J. org. Chem.* 27 (1962) 3674

[45] HLAVKA, J. J. and KRAZINSKI, H. *J. org. Chem.* 28 (1963) 1422

[46] SPENCER, J. L., HLAVKA, J. J., PETISI, J., KRAZINSKI, H. and BOOTHE, J. H. *J. med. Chem.* 6 (1963) 475

[47] STEPHENS, C. R., BEEREBOOM, J. J., RENNHARD, H. H., GORDON, P. N., MURAI, K., BLACKWOOD, R. K. and VON WITTENAU, M. S. *J. Amer. chem. Soc.* 85 (1963) 2643

[48] RITTER, J. J. and MINIERI, P. *J. Amer. chem. Soc.* 70 (1948) 4045

[49] BLACKWOOD, R. K., BEEREBOOM, J. J., RENNHARD, H. H., VON WITTENAU, M. S. and STEPHENS, C. R. *J. Amer. chem. Soc.* 85 (1963) 3943

[50] VON WITTENAU, M. S., BEEREBOOM, J. J., BLACKWOOD, R. K. and STEPHENS, C. R. *J. Amer. chem. Soc.* 84 (1962) 2647

[51] SCOTT, A. I. and BEDFORD, C. T. *J. Amer. chem. Soc.* 84 (1962) 2271

[52] VON WITTENAU, M. S. *J. org. Chem.* 29 (1964) 2746

[53] BEDFORD, C. T. *Ph.D. Thesis*, University of Glasgow, 1963

[53a] YALPANI, M. *Ph.D. Thesis*, University of British Columbia, 1965

[54] GREEN, A., WILKINSON, R. G. and BOOTHE, J. H. *J. Amer. chem. Soc.* 82 (1960) 3946

[55] GREEN, A. and BOOTHE, J. H. *J. Amer. chem. Soc.* 82 (1960) 3950

[56] MUXFELDT, H., BUHR, G. and BANGAT, R. *Angew Chem.* (Int. edn) 1 (1962) 157

[57] BLACKWOOD, R. K., RENNHARD, H. H. and STEPHENS, C. R. *J. Amer. chem. Soc.* 82 (1960) 5194

[58] MUXFELDT, H. and KREUTZER, A. *Naturwissenschaften* 46 (1959) 204; idem *Chem. Ber.* 94 (1961) 881

[59] HOLMLUND, C. E., ANDRES, W. W. and SHAY, A. J. *J. Amer. chem. Soc.* 81 (1959) 4748

[60] HOLMLUND, C. E., ANDRES, W. W. and SHAY, A. J. *ibid.* 81 (1959) 4750

[61] BIRCH, A. J. *Proc. chem. Soc. Lond.* 3 (1962); idem *Fortschr. Chem. org. Nat. Stoffe.* 14 (1957) 186

[62] ROBINSON, R. *Structural Relations of Natural Products*, p. 58. Oxford University Press, London, 1955

[63] BIRCH, A. J., SNELL, J. F. and THOMSON, P. J. *J. chem. Soc.* 425 (1962)

[64] GATENBECK, S. *Biochem. biophys. Res. Comm.* 6 (1961) 422

[65] RICKARDS, R. W. *Chem. Ind. (Lond.)* 1038 (1963)

[66] McCORMICK, J. R. D., JOHNSON, S. and SJOLANDER, N. O. *J. Amer. chem. Soc.* 85 (1963) 1692

[67] McCORMICK, J. R. D., REICHENTHAL, J., JOHNSON, S. and SJOLANDER, N. O. *J. Amer. chem. Soc.* 85 (1963) 1694

[68] McCORMICK, J. R. D., MILLER, P. A., JOHNSON, S., ARNOLD, N. and SJOLANDER, N. O. *J. Amer. chem. Soc.* 84 (1962) 3023

[69] McCORMICK, J. R. D., SJOLANDER, N. O., MILLER, P. A., HIRSCH, U., ARNOLD, N. and DOERSCHUK, A. P. *J. Amer. chem. Soc.* 80 (1958) 6460

[70] MILLER, P. A., HASH, J. H., LINCKS, M. and BOHONOS, N. *Biochem. Biophys. Res. Comm.* 18 (1965) 325

[71] MILLER, P. A., SATUMELLI, A., MARTIN, J. H., MITSCHER, L. A. and BOHONOS, N. *Biochem. Biophys. Res. Comm.* 16 (1964) 285

[72] BOHONOS, N. *Experientia.* 21 (1965) 162

[73] McCORMICK, J. R. D. Personal communication. *J. Amer. chem. Soc.* 1965

[74] DOERSCHUK, A. P., McCORMICK, J. R. D., GOODMAN, J. J., SZUMSKI, S. A., GROWICH, J. A., MILLER, P. A., BITLER, B. A., JENSEN, E. R., MATRISHIN, M., PETTY, M. A. and PHELPS, A. S. *J. Amer. chem. Soc.* 81 (1959) 3069

[75] MUXFELDT, H. and ROGALSKI, W. *J. Amer. chem. Soc.* 87 (1965) 933

[76] CONOVER, L. H., BUTLER, K., JOHNSTON, J. D., KORST, J. J. and WOODWARD, R. B. *J. Amer. chem. Soc.* 84 (1962) 3222

[77] WOODWARD, R. B. In *Chemistry of Natural Products.* Butterworths, London, 1963

[78] FIELDS, T. L., KENDE, A. S. and BOOTHE, J. H. *J. Amer. chem. Soc.* 82 (1960) 1250; 3 (1961) 4612

[79] ARBUZOV, Y. A., BERLIN, Y. U., VOLKOV, Y. P., KOLOSOV, M. N., OVCHINNIKOV, Y. A., HSIEH, Y-Y., TAO, C. and SHEMYAKIN, M. M. *Antibiotiki.* 6 (1961) 585

[80] BROCKMANN, H., BROCKMANN, H. jun., GORDON, J. J., KELLER-SCHLIERLEIN, W., LENK, W., OLLIS, W. D., PRELOZ, V. and SUTHERLAND, I. O. *Tetrahedron Ltrs* 8 (1960) 25

[81] MIYAMOTO, M., MORITA, K., KAWAMATSU, Y., SASAI, M., NOHARA, A., TANAKA, K., TATSUOKA, S., NAKANISHI, K., NAKADAIRA, Y. and BHACCA, N. S. *Tetrahedron Ltrs* 2367 (1964)

[82] SCOTT, A. I., McCAPRA, F. and BHACCA, N. S. *Tetrahedron Ltrs* 869 (1964); SHOPPEE, C. W., RITCHIE, E., TAYLOR, W. C. and MIRRINGTON, R. S. *ibid.* p. 653

[83] MUXFELDT, H., ROGALSKI, W. and STRIEGLER, K. *Chem. Ber.* 95 (1962) 2581

[84] McCORMICK, J. R. D. In *Biogenesis of Antibiotic Substances* (ed. Z. Vanek and Z. Hostalek). Academic Press, London, 1965, Chap. 8

SALAMANDER ALKALOIDS

G. Habermehl

STRUCTURE ELUCIDATION OF THE ALKALOIDS	37
Alkaloids with an Oxazolidine System	37
Alkaloids without an Oxazolidine System	43
Neutral Substances in the Skin Gland Secretion	45
BIOSYNTHESIS	45

IT HAS been known for a long time that the black and yellow spotted fire salamander (*Salamandra maculosa* Laur.) is venomous. The physician Laurentius found in 1768 that the skin gland secretion was the source of toxicity, and Zalesky in 1866 found that the venom has the character of an alkaloid. He isolated an amorphous base which he named *samandarine*[1]. Investigations on the toxicology of the salamander venom by Gessner indicated that it affects the central nervous system. Death occurs by a primary respiratory paralysis. The lethal dose of samandarine is: for the frog 0·019 g, for the mouse 0·0034 g, and for the rabbit 0·001 g/kg of weight[2]. Salamander venom is toxic for all animals, for fishes as well as for amphibia, birds and mammalia. Even the salamander succumbs to its own venom if it enters the blood-stream[1].

Chemical investigations of salamander venom, first by Faust[19], and then from about 1930 by Schöpf and co-workers, showed that the skin gland secretion contains about 10 per cent of a mixture of alkaloids, the rest being proteins and water. The quantity of alkaloids is about 40–50 mg/animal. The best method of winning the secretion is as follows[3]. The animals are narcotized with gaseous carbon dioxide, and then the skin glands are sucked out by means of a glass tube connected to a water pump. The crude secretion stiffens to a gum-like mass, which is ground with sand. From this the alkaloids are extracted with ethanol or isopropanol. On evaporation *in vacuo* a nearly colourless syrup remains[4], which is dissolved in dilute acid, basified with concentrated ammonia and extracted with ether. Then, after addition of sodium hydroxide, the aqueous phase is extracted with methylene chloride. By this procedure one minor alkaloid, cycloneosamandione, which is sparingly soluble in ether, can be isolated.

The paper chromatogram of the crude alkaloid mixture shows eight spots (*cf. Table I*). The two main alkaloids, samandarine and samandarone, can be isolated by precipitation of the sparingly soluble hydrochloride, and the ether insoluble semicarbazone. The minor alkaloid samandaridine is isolated as the water-insoluble sulphate[3, 5]. The remaining minor alkaloids can be separated by Craig-distribution between chloroform and dilute acetic acid[6] or by means of preparative layer chromatography on silica gel[7].

There is a difference in the composition of the alkaloid mixtures from the two subspecies *Salamandra maculosa maculosa* and *S. maculosa taeniata*[4]. The taeniata-form, indigenous to western Europe, has samandarine as the main alkaloid; this is lacking in the maculosa-form indigenous to south-eastern Europe (the main alkaloid there being samandarone).

An outline of the alkaloids is given in *Table I*.

Table I

Name	Formula	m.p.	R_F*	Functional groups		
Alkaloids with oxazolidine-system						
Samandarine	$C_{19}H_{31}NO_2$	188°	0·42	—NH—	—O—	>CHOH
Samandarone	$C_{19}H_{29}NO_2$	190°	0·52	—NH—	—O—	>CO
Samandaridine	$C_{21}H_{31}NO_3$	290°	0·20	—NH—	—O—	γ-lactone
O-Acetyl-samandarine	$C_{21}H_{33}NO_3$	159°	0·55	—NH—	—O—	>CH—O—CO—CH₃
Samandenone	$C_{22}H_{31}NO_2$	191°	0·35	—NH—	—O—	>C=C—C=O
Samandinine	$C_{24}H_{39}NO_3$	170°	0·68	—NH—	—O—	>CH—O—CO—CH₃
Alkaloids without oxazolidine-system						
Cycloneosamandione	$C_{19}H_{29}NO_2$	119°	0·08	>N—C—OH		>C=O
Cycloneosamandaridine	$C_{21}H_{31}NO_3$	282°	0·75	>N—C—OH		γ-lactone
Samanine	$C_{19}H_{33}NO$	197°		—NH—		—CHOH

*Paper: SiO₂-paper Schleicher & Schüll, No. 289.
Solvent: Cyclohexane-Diethylamine 9: 1.

STRUCTURE ELUCIDATION OF THE ALKALOIDS

Alkaloids with an Oxazolidine System

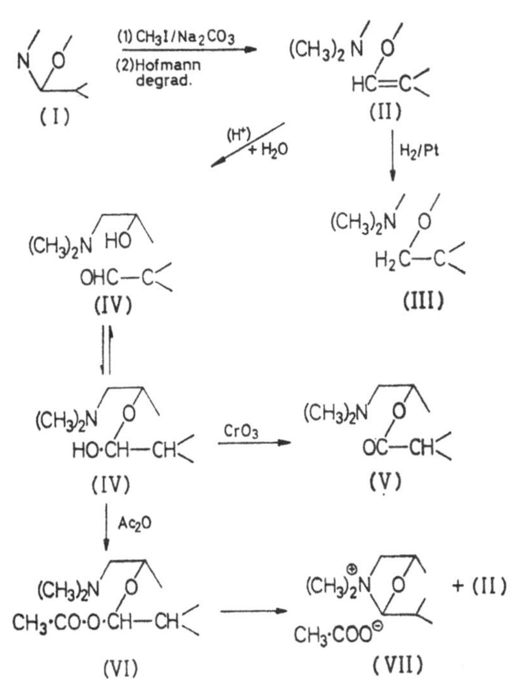

Samandarine—Samandarine, $C_{19}H_{31}NO_2$, is a saturated, secondary amine with a secondary hydroxyl group which can be oxidized to yield the corresponding ketone samandarone. The second oxygen atom, which cannot be detected directly, occurs in an ether bridge. The determination of C-methyl groups proved the presence of two methyl groups. The first insight into the structure of the skeleton of these two alkaloids came from the Hofmann degradation of N-methylsaman-darine-methiodide[3, 5]. The des-base (II) obtained from this reaction contains all the carbon-atoms of the methiodide, proving that the nitrogen atom is present in a ring. Catalytic hydrogenation of the des-base (II) yielded the dihydro-des-base (III). Whereas the des-base is stable towards alkali, on warming with dilute sulphuric acid it adds one molecule of water forming the oxy-dihydro-des-base (IV), a reaction characteristic of enol ethers. This product, on oxidation with chromic acid, reacts as an inner semi-acetal, forming the lactone samandesone (V). These reactions reveal that the nitrogen atom is attached to a carbon atom which is bonded to the ether-oxygen atom or which is adjacent to the carbon atom carrying the oxygen atom. A decision was made on the basis of the following experiment. On warming the oxydihydro-des-base (IV) with acetic anhydride besides the des-base, the quaternary acetate (VII) of the starting material is formed. This may be explained by alkylation of the tertiary nitrogen atom by the first formed acetate (VI). From this it follows that the nitrogen atom and the ether oxygen atom are attached to the same carbon atom.

The enlargement of partial formula I to formula XI results from the following reactions. Samandarine reacts with lithium aluminium hydride to form samandiol[8] (VIII). While samandarine is stable towards lead tetraacetate, samandiol takes up 1 mole of the reagent with formation of 1 mole of formaldehyde. From this it follows that the new hydroxyl group of samandiol must belong to a —NH—CH$_2$—CHOH— grouping as samandiol does not possess a primary hydroxyl group. In this reaction the Schiff base (IX) which is first formed eliminates 1 mole of formaldehyde and undergoes ring closure to (X).

Combination of all these results finally leads to partial formula VIII for samandiol and XI for samandarine.

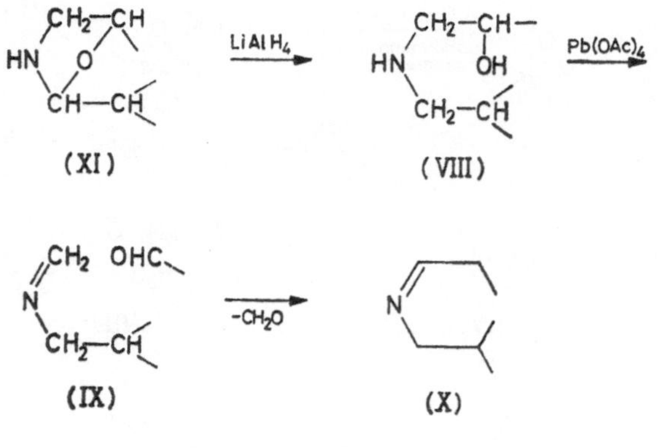

Insight into the skeleton of samandarine was given by dehydrogenation with selenium of samandiol at 320°–340°. In this reaction[9] an oily mixture of hydrocarbons was formed which was separated by means of the addition compounds with trinitrobenzene and with trinitrofluorenone. As the main product there was isolated a crystalline compound, C$_{15}$H$_{16}$, the ultra-violet spectrum of which showed it to be 1,2-dimethyl-5,6-cyclopentenonaphthalene (XII). This hydrocarbon was synthesized as follows:

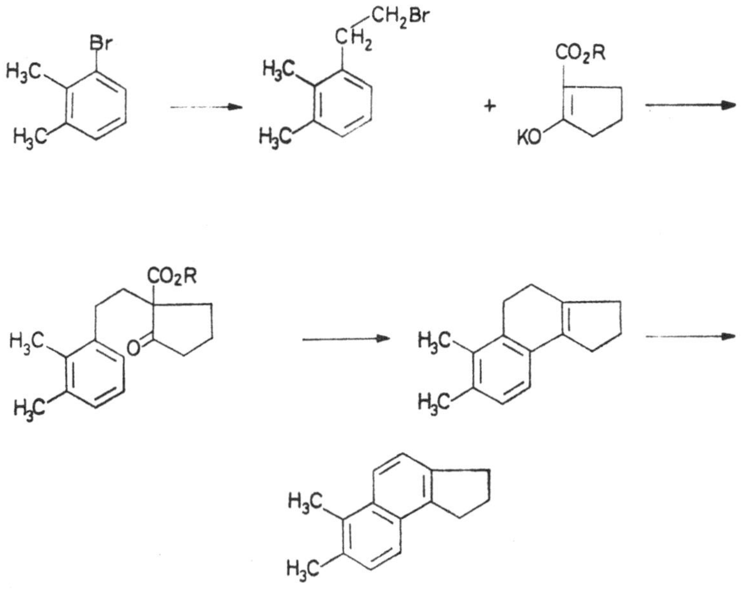

(XII)

The synthetic compound was identical with the product from de-hydrogenation of samandiol.

From the partial formulae XI and XII one could deduce for saman-darine a steroid skeleton with ring A enlarged by a secondary amino group. On this basis there are possibly three formulae (XIII, XIV, XV), in which the position of the secondary hydroxyl group is still uncertain. The infra-red spectrum of samandarone, however, shows that the keto group formed by oxidation of the hydroxyl group is in a five-membered ring (band at 1740 cm^{-1}).

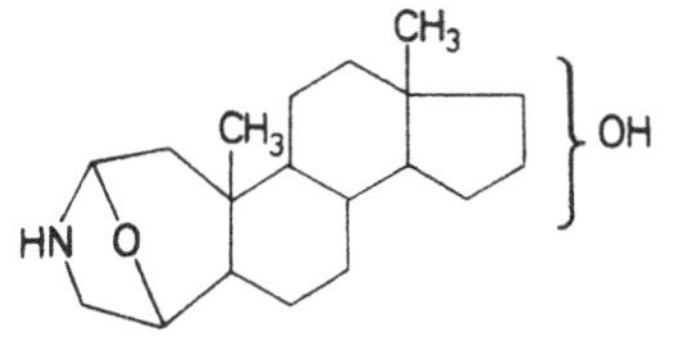

(XIII)

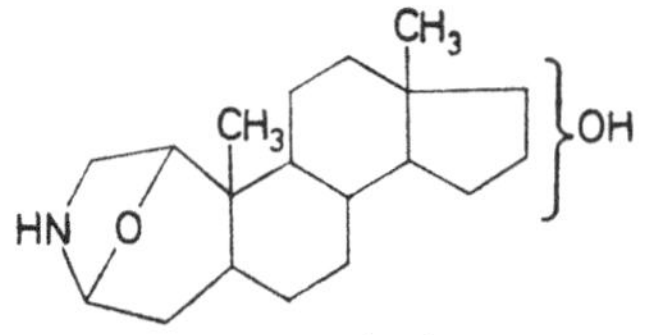

(XV)

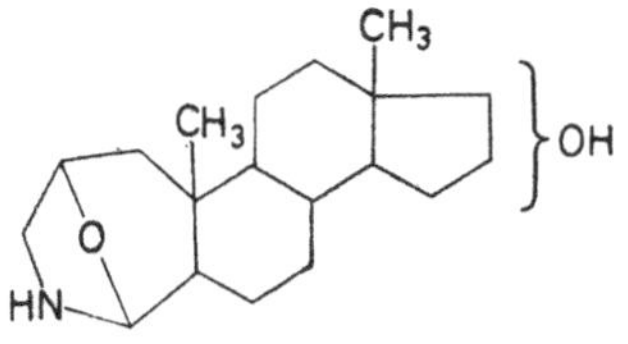

(XIV)

39

The final decision between the three formulae mentioned above was made by X-ray analysis of samandarine-hydrobromide[10], which was performed by the heavy atom method. Samandarine-hydrobromide crystallizes from methanol in orthorhombic prisms with the cell constants a = 12·98 Å; b = 6·28 Å; c = 12·43 Å; β = 95°. The space group is P 2$_1$ with 2 molecules per unit cell. Two- and three-dimensional electron density calculations showed formula XVI (corresponding to XV) to be the correct one. From this it is clear that samandarine has the same configuration as the cholic acids. The secondary hydroxyl group is attached to C-16. The ring A of the steroid skeleton is enlarged by a secondary amino group, which together with carbon atom-2 and with the ether bridge between C-1 and C-4a forms an oxazolidine system.

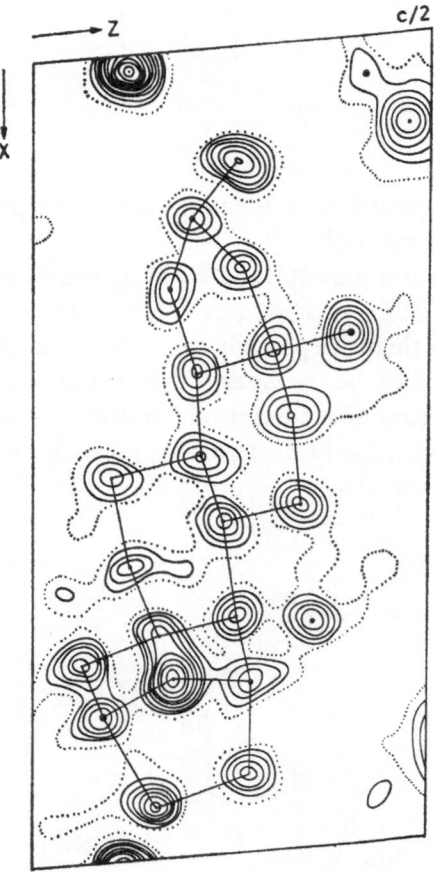

Figure 1. Electron density projection of samandarine-hydrobromide along the b-axis

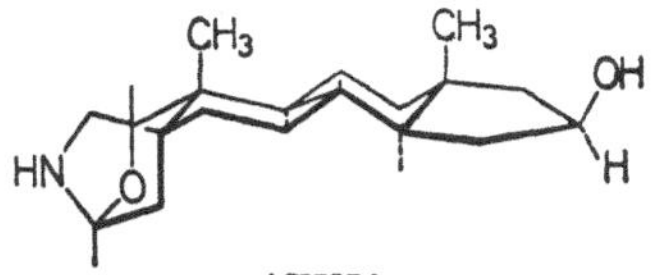

(XVI)

Samandarone—As mentioned above, samandarone (XVII) is the ketone corresponding to samandarine. In *S. maculosa taeniata* it is one of the minor alkaloids, in *S. maculosa maculosa* it is the main alkaloid. It can easily be obtained by oxidation of samandarine with chromic acid, and is reduced stereospecifically to samandarine[5] by sodium and alcohol. From its optical rotatory dispersion curve, which shows a high negative cotton effect ($-7000°$) at 315 mμ, its absolute configuration, and that of samandarine, was shown to be in accord with that of the other natural steroids[11].

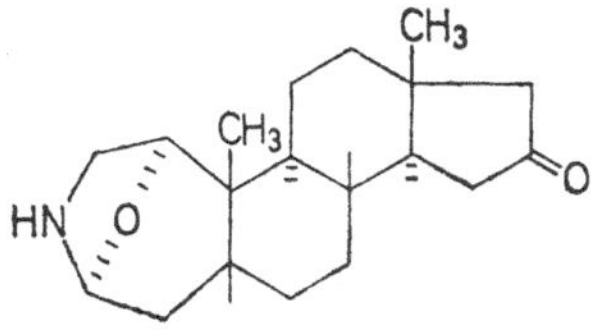

(XVII)

O-Acetyl-samandarine—The structure of O-acetyl-samandarine (XVIII) was elucidated[4] from the infra-red spectrum and from chemical investigations. The infra-red spectrum proved it to be an ester of acetic acid, and it gave samandarine on saponification. A comparison of the natural and the synthesized alkaloid showed the two substances to be identical in all respects.

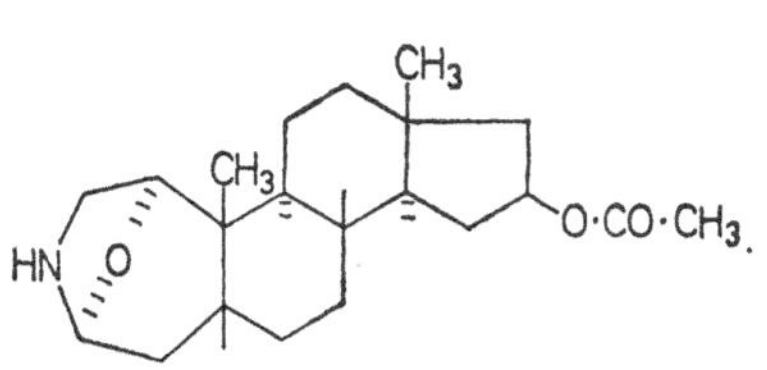

(XVIII)

Samandaridine—Samandaridine (XIX)[12], like the other salamander alkaloids, is a secondary amine. From chemical investigations and from the infra-red spectrum it is seen to possess a five-membered lactone ring. Its structure has also been elucidated by X-ray analysis of the hydrobromide[13]. *Figure 2* shows the electron density map from which its structure results.

4

41

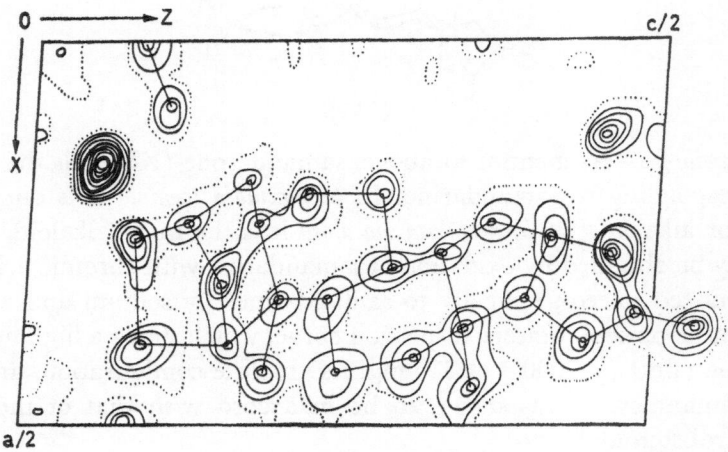

Figure 2. Fourier synthesis of samandaridine hydrobromide; projection along the short axis

The absolute configuration of samandaridine has been determined by partial synthesis[14]. Condensation of N-acetyl-samandarone (XX) with glyoxylic acid yields a mixture of the two carboxylic acids XXI and XXII. Reduction of XXI with sodium borohydride and catalytic hydrogenation proceed stereospecifically to give samandaric acid (XXIII) which on boiling with dilute acid forms samandaridine (XIX).

42

Samandenone—Samandenone[15] contains an oxazolidine system, too, as appears from the infra-red spectrum. Elementary analysis and mass spectrum prove the formula $C_{22}H_{33}NO_2$ to be correct. From infra-red, ultraviolet, NMR- and mass spectra, formula XXIV is determined.

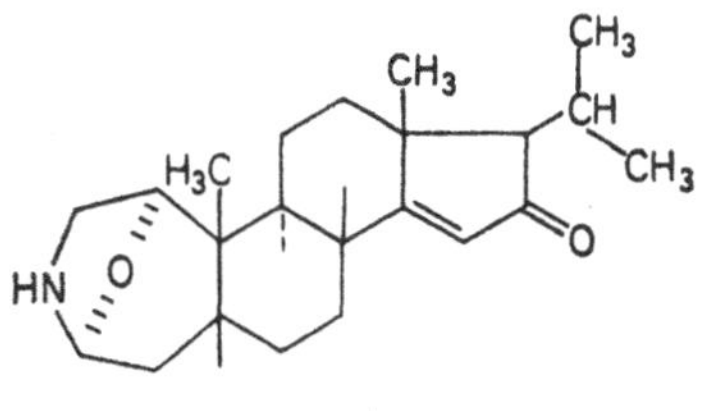

(XXIV)

Samandinine—The structure of this base results from infra-red and mass spectral data.[21]

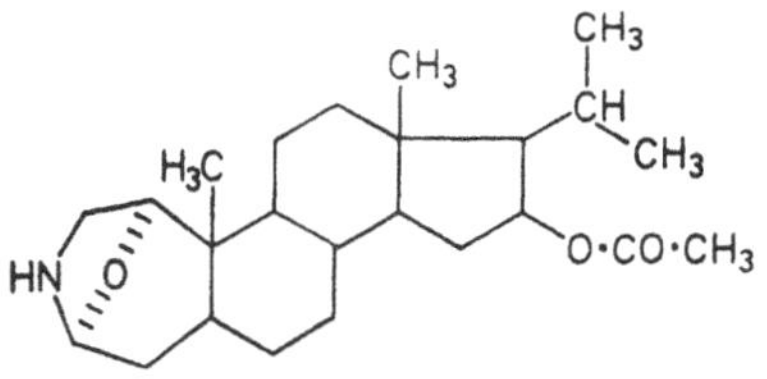

Alkaloids without an Oxazolidine System
Cycloneosamandione—Cycloneosamandione[6] is an isomer of samandarone but lacks the oxazolidine system. From its infra-red spectrum it is evident that one oxygen atom belongs to a keto group in a five-membered ring. The second one is incorporated in an aldehyde-ammonia grouping which can react in the two tautomeric forms

$$\begin{array}{cc}\diagdown \ \ \diagup \\ N-C-OH \ \ \text{and} \ \ \diagup \ \diagup \\ \diagup \ \ \diagdown \end{array} \quad NH \ O{=}C \quad$$ the first of which is distinguished

by the prefix 'cyclo' in the name of the compound.

Acetylation of cycloneosamandione yields a neutral mono-acetyl-compound, the infra-red spectrum of which shows the bands for an aldehyde group at 1720 cm.$^{-1}$ and at 2700 cm.$^{-2}$. The NMR spectrum, too, gives information about this carbonyl group. While the NMR spectrum of samandarone shows two signals at 9·1 and 9·2τ for the angular C-18 and C-19 methyl groups, in the NMR spectrum of cycloneosamandione only the signal at 9·1τ for the C-18 methyl group is present. Additionally there are found two signals for the two protons of an > CHOH group, at 5·5τ for CH and at 3·25τ for OH. The NMR spectrum of N-acetyl-neosamandione, too, shows—like that of cycloneosamandione—the signal for the C-18 methyl group. Instead

43

of the two signals of the > CHOH group, however, one finds a signal at 0.09τ for the proton of an aldehyde group. This makes it evident that the C-19 methyl group is replaced by the aldehyde group[16]. The stereochemistry of cycloneosamandione, however, is not identical with that of the alkaloids mentioned so far. The oxygen-free degradation product, neosamane, prepared by Wolff–Kishner–Huang–Minlon reduction of cycloneosamandione is not identical with 3-aza-A-homo-5α- or 5β-androstane respectively, prepared by Shoppee and Krueger[20].

The structure of the alkaloid, therefore, was finally elucidated by X-ray analysis. Two- and three-dimensional electron density calculations proved the correct formula of cycloneosamandione to be **XXV**[16].

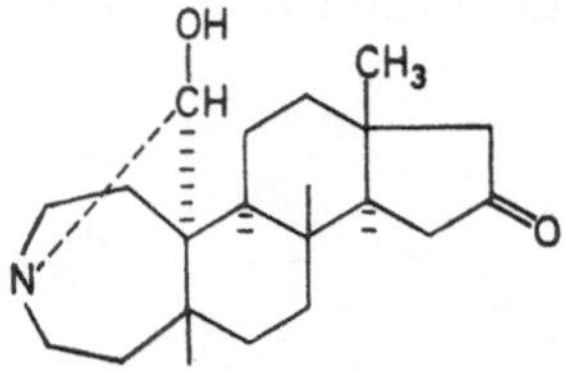

(XXV)

Cycloneosamandione thus has the same basic structure as samandarone with the one, but essential difference of having the aldehyde group in the α configuration in place of the β-methyl group. The arrangement of the rings C and D as well as the absolute configuration were elucidated by the optical rotatory dispersion curve of cycloneosamandione which is quite identical with that of samandarone.

Cycloneosamandaridine—A second minor alkaloid without the oxazolidine system is cycloneosamandaridine, which was isolated in very small amounts. From its infra-red spectrum and that of its N-acetyl derivative it was seen to possess an aldehyde ammonia grouping as well as a γ-lactone ring. From the mass spectrum results the formula $C_{21}H_{33}NO_3$. From this it is obvious that cycloneosamandaridine (**XXVI**) unites the properties of cycloneosamandione with those of samandaridine.

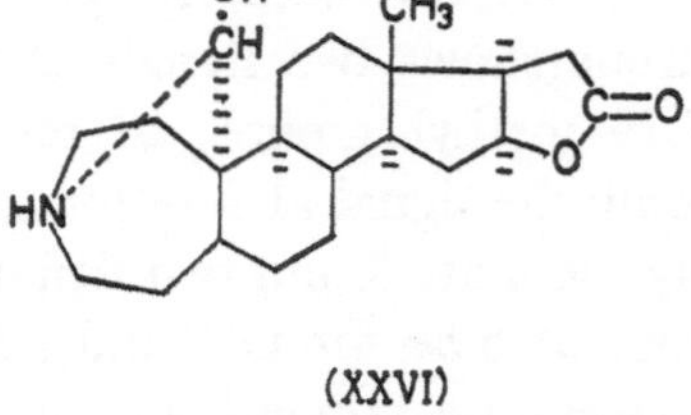

(XXVI)

Samanine—The structure of this minor alkaloid was elucidated from infra-red and mass spectral data of the base as well as its O, N-di-acetyl-derivatives.[21]

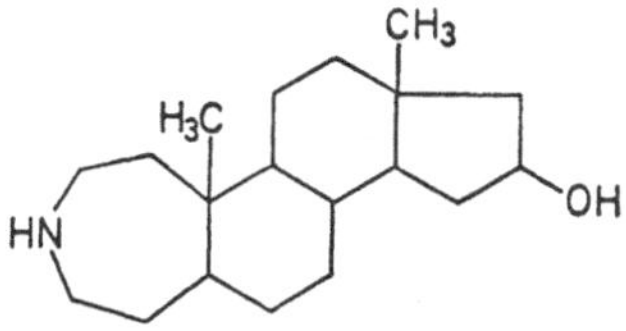

Neutral Substances in the Skin Gland Secretion
Besides the alkaloids, a fraction of neutral, lipophilic substances can be isolated. By column chromatography on silica gel six components have been isolated, and the structures of four of them have been elucidated [18]. The main part consisted of cholesterol; the other compounds were esters of cholesterol, namely the stearate, palmitate and the oleate.

BIOSYNTHESIS

It is well known today that cholesterol is a key substance in steroid metabolism. Thus, cholesterol is the intermediate in the formation of the steroidal sex hormones. As the skin gland secretion of *Salamandra maculosa* contains remarkable quantities of cholesterol, it seemed reasonable, also, to suppose a steroid precursor for the salamander alkaloids, especially as samandenone and samandaridine are substituted at C-17. They could represent modified intermediate stages in the course of the biosynthesis of samandarine or samandarone. Experiments with ^{14}C-labelled acetate and cholesterol showed that all salamander alkaloids are formed, like other steroids, from acetate via cholesterol.[17] One could suppose that a cholic acid side chain is degraded (by β-oxidation) to a compound (XXVII) with a side chain corresponding to that of samandenone.

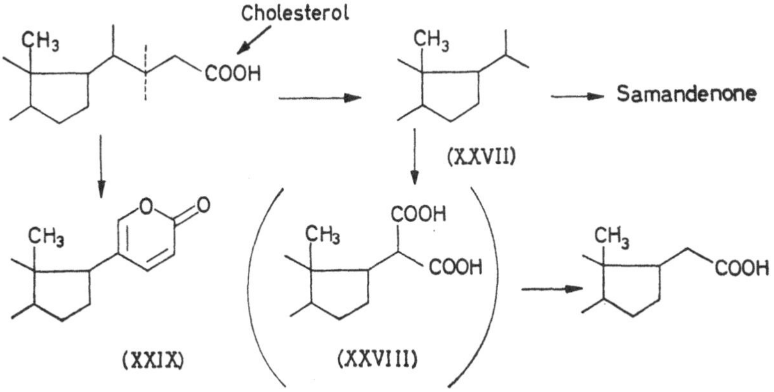

Further oxidation could then lead to malonic acid (XXVIII) by the decarboxylation of which the CH_2-COOH side chain is formed. Ring closure with the hydroxyl group at C-16 then gives the lactone samandaridine.

Similarly, in the skin glands of *Bufo bufo*, which is nearly related to *Salamandra maculosa*, bufotaline (XXIX) could be formed from the cholic acid precursor without loss of carbon atoms[13].

For the anomalous configuration at C-10 of cycloneosamandione there is an explanation, too. The ring enlargement of ring A of all salamander alkaloids must follow a cleavage between the carbon atoms 2 and 3 and insertion of the nitrogen atom. If there is a carbonyl function in the opened compound at C-3 (XXX) the A/B ring system could now undergo a reverse Michael cleavage to (XXXI). Now at the carbon atom 10 an epimerization (XXXII) can take place. After the insertion of nitrogen and following ring closure to (XXXIII) the aldehyde group has the α-configuration. This reaction is never observed with other steroids containing a C-19 aldehyde group as the decalin system XXXIV shows no tendency to be converted into the cyclodecane system XXXV.

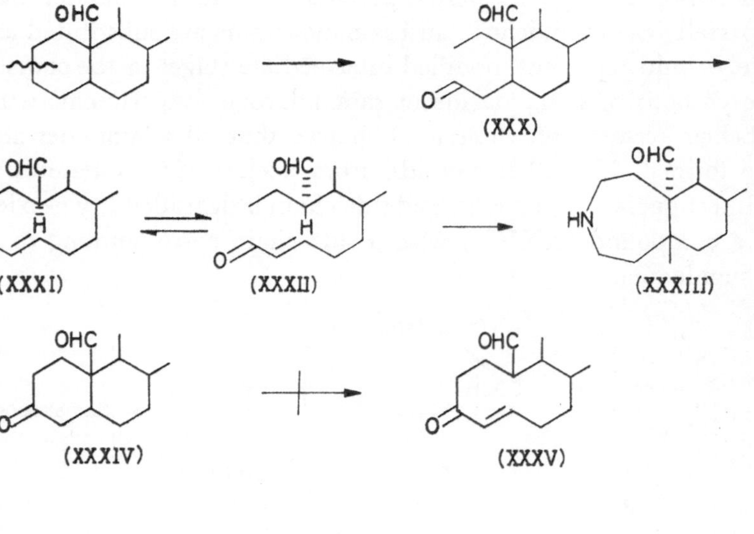

REFERENCES

[1] ZALESKY, S. *Med. chem. Unters.* Hoppe-Seyler 1 (1866) 85
[2] GESSNER, O. 'Tierische Gifte', *Handb. exp. Path. Erg. Werk.* Vol. 6, p. 56
[a] GESSNER, O. Habilitationsschrift Marburg: *Ber. Ges. Bef. ges. Naturwiss.* Marburg 61 (1926) 138

[b] GESSNER, O. and MÖLLENHOFF, F. *Arch. exp. Path. Pharmak.* *167,* 638 (1932)

[c] GESSNER, O. and ESSER, W. *ibid.* 179 (1935) 639

[d] GESSNER, O. and URBAN, G. *ibid.* 187 (1937) 378

[3] SCHÖPF, C. and BRAUN, W. *Liebigs Ann.* 514 (1934) 69

[4] HABERMEHL, G. *Liebigs Ann.* 679 (1964) 164

[5] SCHÖPF, C. and KOCH, K. *Liebigs Ann.* 552 (1942) 37

[6] SCHÖPF, C. and MÜLLER, O. W. *Liebigs Ann.* 633 (1960) 127

[7] HABERMEHL, G. and VOGEL, G. *Chem. Ber.* 98 (1965)

[8] SCHÖPF, C., BLÖDORN, H. K., KLEIN, D. and SEITZ, G. *Chem. Ber.* 83 (1950) 372

[9] SCHÖPF, C. and KLEIN, D. *Chem. Ber.* 87 (1954) 1638

[10] SCHÖPF, C. *Experientia.* 17 (1961) 285

[11] WÖLFEL, E., SCHÖPF, C., WEITZ, G. and HABERMEHL, G. *Chem. Ber.* 94 (1961) 2361

[12] SCHÖPF, C. and KOCH, K. *Liebigs Ann.* 552 (1942) 62

[13] HABERMEHL, G. *Chem. Ber.* 96 (1963) 143

[14] HABERMEHL, G. *Chem. Ber.* 96 (1963) 840

[15] HABERMEHL, G. *Chem. Ber.* 98 (1965)

[16] HABERMEHL G. and GÖTTLICHER, S. *Chem. Ber.* 98 (1965) 1

[17] HABERMEHL, G. and HAAF, A. *Chem. Ber.* 100 (1967), in press

[18] HABERMEHL, G. *Liebigs Ann.* 680 (1964) 104

[19] FAUST, S. *Arch. exp. Path. Pharmak.* 41 (1898) 229; *ibid.* 43 (1900) 84

[20] SHOPPEE, C. W. and KRUEGER, G. *J. chem. Soc., Lond.* (1961) 3641

[21] HABERMEHL, G. To be published

3

ELECTROPHILIC MOLECULAR REARRANGEMENTS*

T. S. Stevens

INTRODUCTION	48
REARRANGEMENTS INVOLVING 1 → 2 MIGRATION	49
Ammonium Nitrogen to Anionic Carbon	49
Other 'Onium Atoms to Anionic Carbon	54
Oxygen to Anionic Carbon	55
Uncharged Nitrogen to Anionic Carbon	58
Some General Considerations	59
Sulphur to Anionic Carbon	61
Uncharged Carbon to Anionic Carbon	61
'Onium Sulphur to Anionic Nitrogen	62
Ammonium Nitrogen to Anionic Oxygen	63
MIGRATIONS OVER LONGER RANGE	64
The Smiles Rearrangement and Related Changes	64
Rearrangement of Imino-Ethers and Amidines	67
Miscellaneous Rearrangements	68

INTRODUCTION

AN ELECTROPHILIC molecular rearrangement is one in which the migrating radical leaves an electron pair on the atom of departure (migration origin) and becomes linked through an electron pair on the destination atom (migration terminus). An early example is the thermal isomerization of isocyanides to nitriles[1]:

$$: \bar{C} ::: \overset{+}{N} : R \longrightarrow R : C ::: N :$$

The process results in the reduction of one unit of covalency and the loss of one unit of positive formal charge at the migration origin, with concomitant gain of one unit of covalency and of positive formal charge at the migration terminus. In many cases the destination atom needs to exhibit strongly nucleophilic character, and the active lone pair has to be developed by using a powerful base to extract a proton:

* The manuscript for this chapter was submitted in May, 1965.

$$\begin{matrix} Ph \cdot CH_2 \\ \\ Ph \cdot CH_2 \end{matrix}\!\!\!\nearrow\!\!\!\searrow \overset{+}{N}Me_2 \xrightarrow{\;\overset{-}{N}H_2\;} \begin{matrix} Ph \cdot \overset{-}{C}H \\ \\ Ph \cdot CH_2 \end{matrix}\!\!\!\nearrow\!\!\!\searrow \overset{+}{N}Me_2 \longrightarrow \begin{matrix} Ph \cdot CH \cdot NMe_2 \\ | \\ Ph \cdot \overset{-}{C}H_2 \end{matrix}$$

The most significant cases involve a $1 \to 2$ migration of the radical R, from one atom to the adjacent one, but transformations are well known, e.g. the Smiles rearrangement (p. 64), in which the radical migrates over a greater distance.

REARRANGEMENTS INVOLVING $1 \to 2$ MIGRATION
The principal types, with examples, are summarized in *Table I.*

Table I. Rearrangements Involving '1 → 2 Migration

Type	Example	Reagent
$\overset{+}{N} \longrightarrow \overset{-}{C}$	$(Ph \cdot CH_2)_2\overset{+}{N}Me_2 \longrightarrow Ph \cdot CH(NMe_2)CH_2Ph$	$\overset{-}{N}H_2$
$\overset{+}{O} \longrightarrow \overset{-}{C}$	$(Ph \cdot CH_2)_2O \longrightarrow Ph \cdot CH(OH) \cdot CH_2Ph$	PhLi
$N \longrightarrow \overset{-}{C}$	$Ph \cdot CO \cdot CH_2 \cdot \underset{\underset{CH_2Ph}{\vert}}{NPh} \longrightarrow Ph \cdot CO \cdot \underset{\underset{CH_2Ph}{\vert}}{CH} \cdot NHPh$	KOH fusion
$C \longrightarrow \overset{-}{C}$	$Ph_3C \cdot CH_2Cl \longrightarrow Ph_2\overset{-}{C} \cdot CH_2Ph$	Na
$\overset{+}{S} \longrightarrow \overset{-}{N}$	$Ts \cdot \overset{-}{N} \cdot \overset{+}{S}(C_3H_5) \cdot CH_2Ph \longrightarrow TsN(C_3H_5) \cdot S \cdot CH_2Ph$	Heat
$\overset{+}{N} \longrightarrow \overset{-}{O}$	$\overset{-}{O}-\overset{+}{N}MePh \cdot CH_2Ph \longrightarrow Ph \cdot CH_2 \cdot O \cdot NMePh$	Heat

Ammonium Nitrogen to Anionic Carbon

Apart from the isocyanide → nitrile change, these rearrangements require the extraction of a proton by an alkaline reagent to produce the required carbanion. In many cases it is not certain that proton removal and rearrangement are separate, independent processes. This removal is very difficult—requiring a reagent of extreme alkalinity—when no electron-attracting groups, apart from quaternary nitrogen, are present; but just in these cases, the anionic carbon atom, once achieved, has the greatest tendency to be relieved of its charge by electrophilic rearrangement.

The earliest rearrangements of this kind to be studied in detail were those of benzylphenacylammonium ions[2]:

$$Ph \cdot CO \cdot CH_2 \cdot \underset{\underset{\underset{(I)}{CH_2Ph}}{\vert}}{\overset{+}{N}Me_2} \xrightarrow[\overset{-}{O}Et]{\overset{-}{O}H\ or} Ph \cdot CO \cdot \underset{\underset{CH_2Ph}{\vert}}{\overset{-}{C}H} \cdot \overset{+}{N}Me_2 \longrightarrow Ph \cdot CO \cdot \underset{\underset{CH_2Ph}{\vert}}{CH} \cdot NMe_2$$

effected by warming with aqueous or ethanolic alkali. Here the first stage (neutralization of a weak acid) can be recognized independently of the rearrangement. The reaction is first order in ammonium ion and is not accelerated by further molecules of alkali. An Arrhenius activation energy of ~30–33 kcal is indicated by the few observations at different temperatures. Reaction is understandably much faster in solvents of low dielectric constant[3].

A wide range of radicals of benzyl or allyl type will migrate, as will phenacyl, $Ph \cdot CO \cdot CH_2 \cdot$, but simple alkyl and aryl groups do not do so under the usual conditions[4]. Electron-attracting substituents in the benzene ring of the phenacyl group of (I) retard the rearrangement slightly[5], no doubt by stabilizing the negative charge on the intermediate zwitterion ('Ylid') and so diminishing the driving force for the second, rate-determining stage. Similar substituents in the migrating benzyl group have a strong accelerating effect (*Table II*)[3]. The change may be regarded formally as a nucleophilic substitution by $\bar{C}$ of $\overset{+}{N}$ on the benzyl carbon atom. It is then natural that a benzyl or allyl group should migrate more easily than a simple alkyl or aryl, and that electron-attracting substituents should promote the change.

Table II. Migration Velocities of Substituted Benzyl Groups

	Substituent in benzyl group					
	Cl	Br	I	NO$_2$	OMe	Me
o—	36	48	81	1040	1·9	15
m—	2·4	2·1	1·9	3·8	0·93	0·97
p—	2·6	2·8	3·2	73	0·76	1·06

Unsubstituted benzyl = 1

That the reaction is strictly intramolecular is shown by 'mixing experiments' with labelled materials[6]:

$$p\text{-Br} \cdot C_6H_4 \cdot CO \cdot CH_2 \cdot \overset{+}{N}Me_2$$
$$\overset{|}{^{14}CH_2Ph}$$
$$Ph \cdot CO \cdot CH_2 \cdot \overset{+}{N}Me_2$$
$$\overset{|}{CH_2Ph}$$
$$\overset{\bar{O}H}{\longrightarrow}$$
$$Br \cdot C_6H_4 \cdot CO \cdot CH \cdot NMe_2$$
$$\overset{|}{^{14}CH_2Ph}$$
$$Ph \cdot CO \cdot CH \cdot NMe_2$$
$$\overset{|}{CH_2Ph}$$
$$(II)$$

no radioactivity appeared in (II). Further, a radical attached to nitrogen through an asymmetric carbon atom migrates with substantially complete retention of configuration[7, 8]:

50

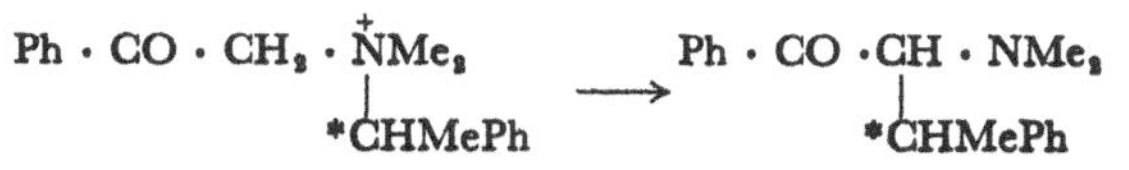

$$\begin{array}{ccc} Ph \cdot CO \cdot CH_2 \cdot \overset{+}{N}Me_2 & & Ph \cdot CO \cdot CH \cdot NMe_2 \\ | & \longrightarrow & | \\ \bullet CHMePh & & \bullet CHMePh \end{array}$$

The migration of a radical of allyl type, $\cdot CH_2 \cdot CH{=}CHR$, may or may not involve 'inversion' to $CH_2{=}CH \cdot CHR \cdot$, as in the following cases[8]:

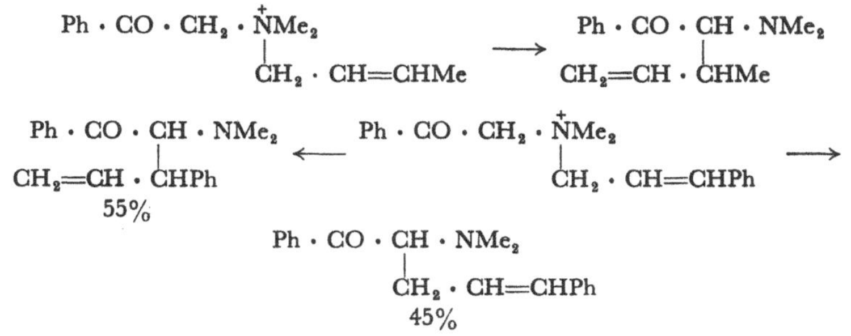

$$\begin{array}{ccc} Ph \cdot CO \cdot CH_2 \cdot \overset{+}{N}Me_2 & & Ph \cdot CO \cdot CH \cdot NMe_2 \\ | & \longrightarrow & | \\ CH_2 \cdot CH{=}CHMe & & CH_2{=}CH \cdot CHMe \end{array}$$

$$\begin{array}{ccc} Ph \cdot CO \cdot CH \cdot NMe_2 & & Ph \cdot CO \cdot CH_2 \cdot \overset{+}{N}Me_2 \\ | & \longleftarrow & | & \longrightarrow \\ CH_2{=}CH \cdot CHPh & & CH_2 \cdot CH{=}CHPh \\ 55\% & & \end{array}$$

$$\begin{array}{c} Ph \cdot CO \cdot CH \cdot NMe_2 \\ | \\ CH_2 \cdot CH{=}CHPh \\ 45\% \end{array}$$

The rates of migration of benzyl, allyl, and cinnamyl are in the approximate ratios $1 : 500 : 10^5$.

Rearrangement of analogous fluorenylammonium ions[9] is effected almost equally easily, under the same conditions:

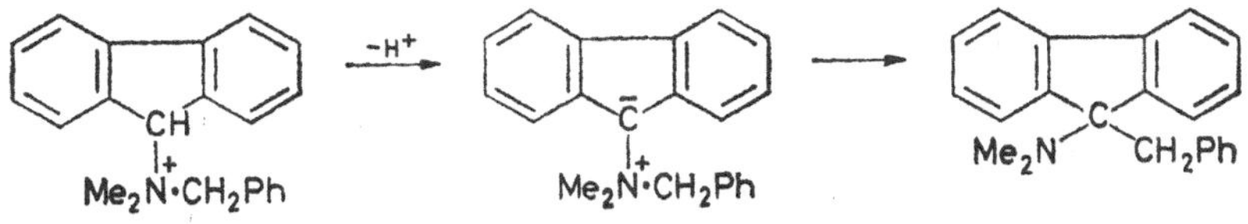

When the migration terminus is less favourably substituted, more energetic conditions are required. The ion (III) rearranges to (IV) under the conditions of a Hofmann degradation[10]; here the intermediate zwitterion is substantially more basic than in the phenacyl series, and also in general more robust chemically, so that migration of a methyl group can be effected prior to gross decomposition.

On the other hand, SOMMELET[11] observed that if the hydroxide corresponding to (III) is kept at a moderate temperature in a desiccator, it undergoes a different kind of rearrangement. The intensely reactive zwitterion (V), formed only in very minor amount, undergoes intramolecular nucleophilic attack on the benzene nucleus, giving finally (VI). By action of the extremely strong base, phenyl-lithium, (III) affords simultaneously (IV) and (VI)[12].

51

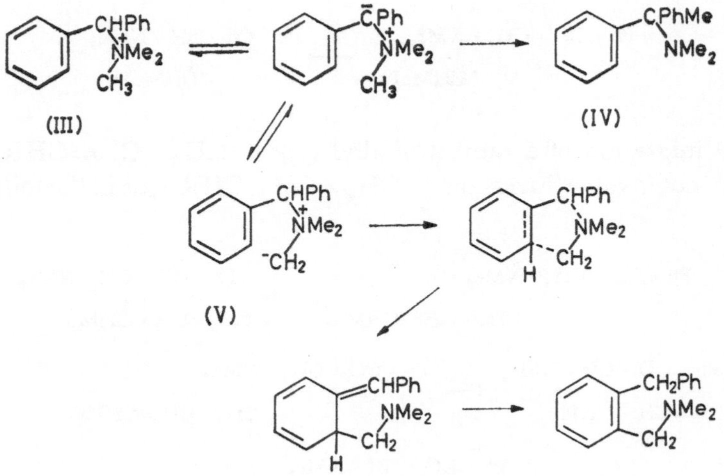

$$\text{(III)} \rightleftharpoons \text{(IV)}$$

$$\text{(V)} \longrightarrow$$

$$\text{(VI)}$$

The dibenzyldimethylammonium ion may be rearranged by fusion with sodamide[4], by treatment with sodamide or potassamide in liquid ammonia[13], or by the action of phenyl-lithium in ether[9], giving the alternative products (VII) and (VIII).

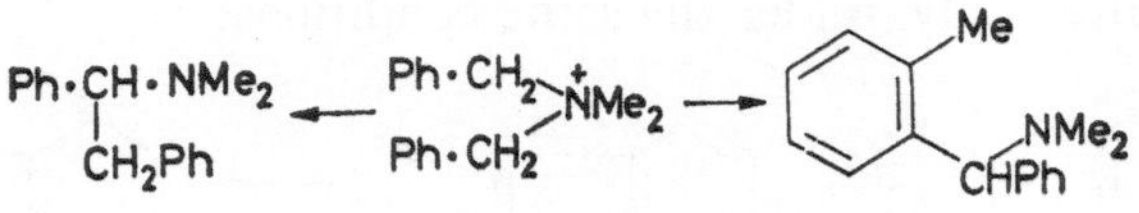

$$\text{Ph·CH·NMe}_2 \quad \longleftarrow \quad \text{Ph·CH}_2\cdot\overset{+}{\text{N}}\text{Me}_2 \quad \longrightarrow$$
$$\quad \overset{|}{\text{CH}_2\text{Ph}} \qquad\qquad \text{Ph·CH}_2$$

(VII)   (VIII)

Rearrangement to (VII) is favoured by high temperature. It is not obvious why the energy of activation should be greater for the production of (VII), but the entropy term should clearly be less, since fewer degrees of freedom are frozen in the transition state for this transformation than in that leading to (VIII)[14].

The fluorenyltrimethylammonium ion does not rearrange[15], no doubt owing to the less basic character of the related zwitterion; but benzyltrimethylammonium ion undergoes the $1 \rightarrow 2$[12] and the Sommelet[16] change:

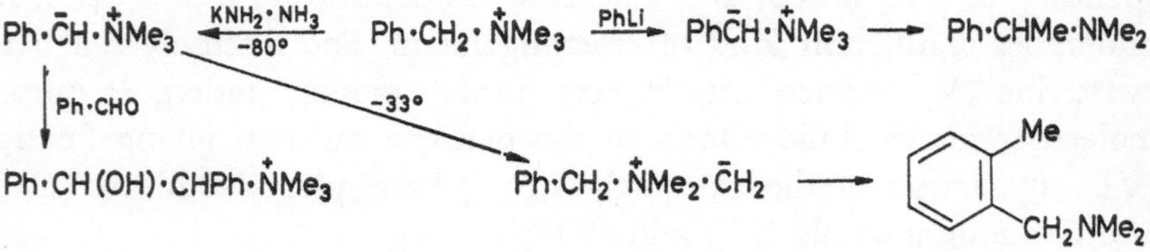

$$\text{Ph·}\overset{-}{\text{C}}\text{H·}\overset{+}{\text{N}}\text{Me}_3 \xleftarrow[-80°]{\text{KNH}_2\cdot\text{NH}_3} \text{Ph·CH}_2\cdot\overset{+}{\text{N}}\text{Me}_3 \xrightarrow{\text{PhLi}} \text{Ph}\overset{-}{\text{C}}\text{H·}\overset{+}{\text{N}}\text{Me}_3 \longrightarrow \text{Ph·CHMe·NMe}_2$$

$$\downarrow \text{Ph·CHO} \qquad\qquad -33°$$

$$\text{Ph·CH(OH)·CHPh·}\overset{+}{\text{N}}\text{Me}_3 \qquad\qquad \text{Ph·CH}_2\cdot\overset{+}{\text{N}}\text{Me}_2\cdot\overset{-}{\text{C}}\text{H}_2 \longrightarrow$$

52

The type of quinonoid intermediate postulated in the reactions leading to (VI) and (VIII) has been isolated in a case in which its further
transformation is blocked [17]:

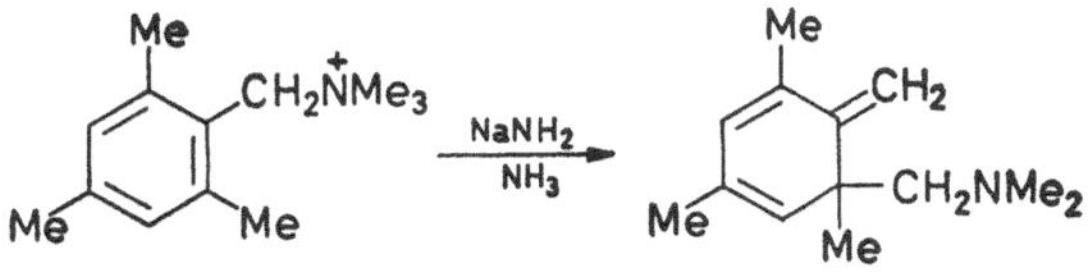

$$\text{Me, CH}_2\overset{+}{\text{N}}\text{Me}_3 \quad\xrightarrow[\text{NH}_3]{\text{NaNH}_2}\quad \text{Me, CH}_2,\ \text{CH}_2\text{NMe}_2$$

WITTIG [18] records the effect of experimental conditions on the transformation of an isoindolinium ion (see *Table III*):

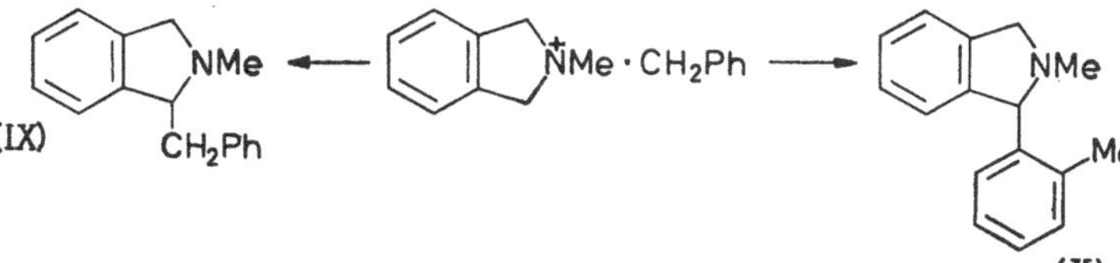

$$\text{(IX)}\quad\overset{\text{NMe}}{\underset{\text{CH}_2\text{Ph}}{}}\quad\longleftarrow\quad \overset{+}{\text{N}}\text{Me}\cdot\text{CH}_2\text{Ph}\quad\longrightarrow\quad \text{NMe} \quad\text{(X)}$$

Under Hofmann conditions a further product is *N*-methyl-isoindoline.

Table III. *Rearrangement of Benzylmethylisoindolinium Ion*

Conditions	Yield of (IX)	Yield of (X)
NaNH$_2$—NH$_3$, —33°	—	87%
PhLi—Et$_2$O, 20–30°	—	69%
NaOEt—EtOH, 80°	44%	—
PhLi—Bu$_2$O, 120°	41	—
Hofmann	33	—

A cyclopropyl system can promote rearrangement in the same way
as an allylic double bond [19]:

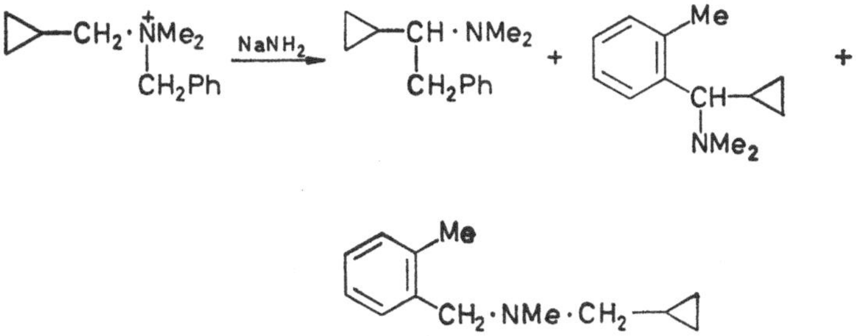

$$\triangleright\text{—CH}_2\cdot\overset{+}{\text{N}}\text{Me}_2 \quad\xrightarrow{\text{NaNH}_2}\quad \triangleright\text{—CH}\cdot\text{NMe}_2 \ + \ \overset{\text{Me}}{\underset{\text{CH—}\triangleleft}{}}\ + $$

$$\overset{\text{Me}}{\underset{\text{CH}_2\cdot\text{NMe}\cdot\text{CH}_2\text{—}\triangleleft}{}}$$

Three intriguing synthetic applications of these changes may be recorded. HAUSER effected two Sommelet sequences[13, 20]—

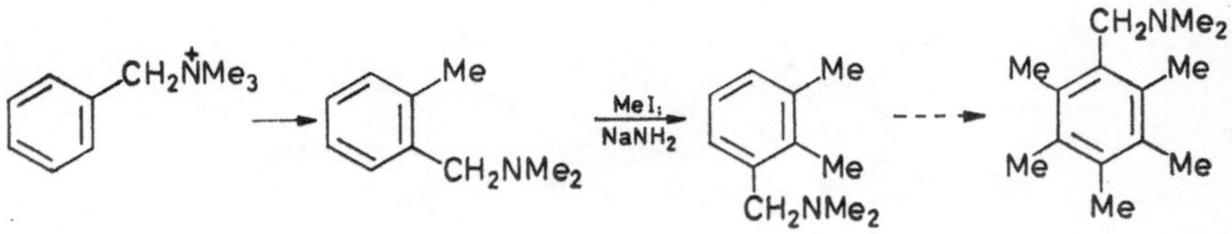

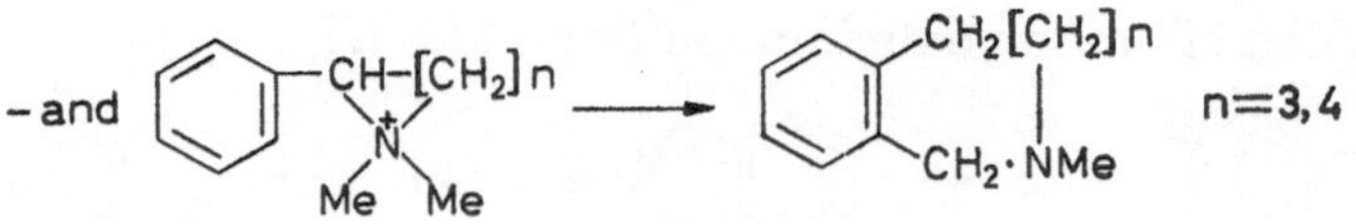

and WITTIG[9] synthesized dibenzocyclooctatetraene:

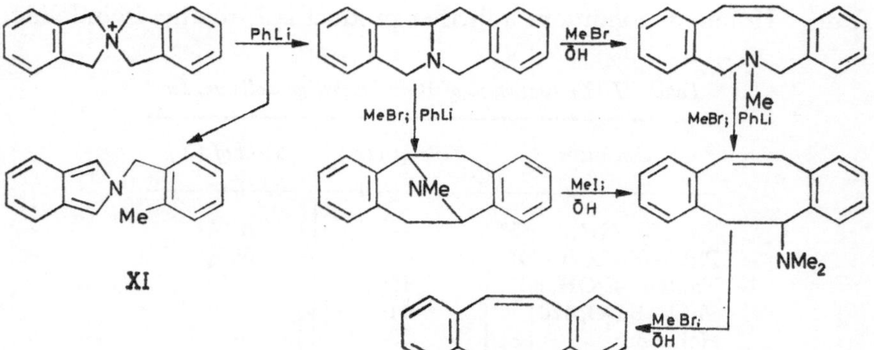

XI

Compound (XI) was the first unreduced isoindole to be reported.

The thermal ($\sim 200°$) isomerization of *isocyanides* to nitriles has been studied in detail by Rabinovitch[21]. The first-order reaction has an activation energy of 34–38 kcal.

Other 'Onium Atoms to Anionic Carbon

Arsonium and stibonium ions can undergo $1 \rightarrow 2$ rearrangement[22]:

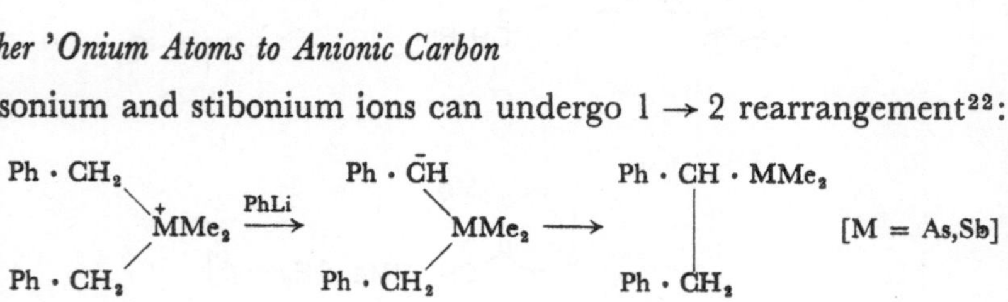

Sulphonium ions show both $1 \rightarrow 2$ rearrangement and the Sommelet change[23, 24, 25]:

$$Ph \cdot CO \cdot CH_2 \cdot \overset{+}{S}Me \cdot CH_2Ph \xrightarrow[\text{MeOH}]{\text{NaOMe}} Ph \cdot CO \cdot \underset{\overset{|}{CH_2Ph}}{CH} \cdot SMe$$

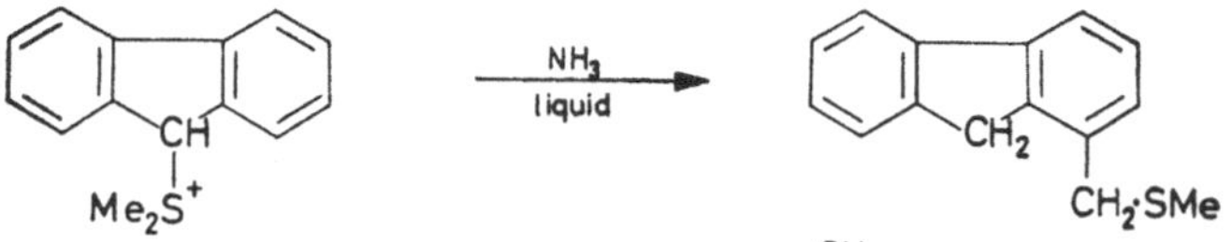

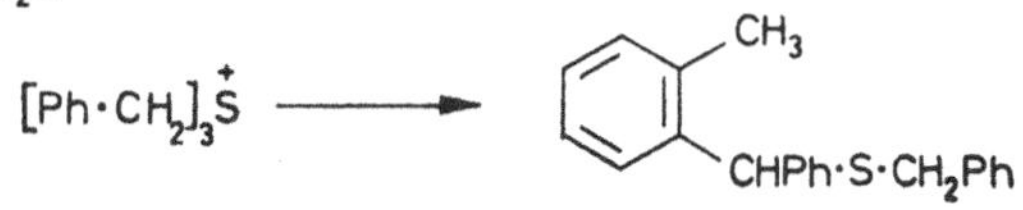

Oxygen to Anionic Carbon

While isolated observations are recorded earlier by SCHORIGIN [26] and by SCHLENK and BERGMANN [27], we are indebted to WITTIG for systematic investigation. In comparison with the rearrangement of quaternary ammonium ions, the initial extraction of a proton is here, as would be expected, substantially more difficult; further the essential change in the status of the oxygen atom (neutral $\rightarrow$ anionic) is rather less favoured than that of the ammonium nitrogen (cationic $\rightarrow$ neutral).

On treatment with phenyl-lithium in ether, fluorenyl ethers undergo rearrangement[28] as shown below. Reaction proceeds at $40°$ for R = allyl or benzyl, at $100°$ for R = methyl, ethyl, or p-nitrophenyl, and fails for R = phenyl—in consonance with the diminishing ease of ordinary nucleophilic substitution on R. The same rearrangement of allyl and benzyl, but not alkyl and aryl fluorenyl ethers is effected by butanolic sodium butoxide at $120°$[29].

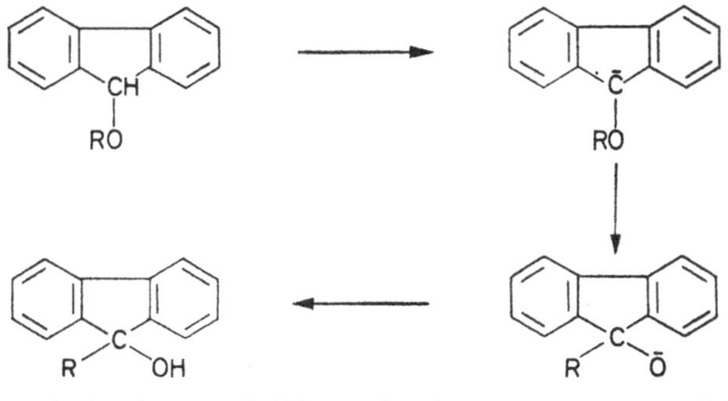

Benzyl methyl ether and dibenzyl ether are rearranged by phenyl-lithium[30], giving $Ph \cdot CHR \cdot OH (R = Me, CH_2Ph)$. Benzyl ethyl ether yields benzyl alcohol and ethylene, a process analogous to the common mode of fission of dialkyl ethers by reactive organo-alkali compounds. Benzyl phenyl ether affords 1,1,2-triphenylethane; conjecturally,

$$PhO \cdot \bar{C}HPh + CH_2Ph \cdot OPh \longrightarrow PhO \cdot CHPh \cdot CH_2Ph + \bar{O}Ph$$

$$\bar{P}h + PhO \cdot CHPh \cdot CH_2Ph \longrightarrow Ph_2CH \cdot CH_2Ph + \bar{O}Ph$$

The reaction has also been formulated[31] in terms of the production of phenylcarbene as intermediate ($Ph\bar{C}HOPh \rightarrow PhCH < + \bar{O}Ph$). The same rearrangements of ethers have been effected[32] by treating them with potassamide in liquid ammonia and refluxing the product in ether.

Benzyl phenacyl ether, heated with butanolic sodium butoxide, evidently undergoes an electrophilic rearrangement, followed by a benzilic change, giving 2,3-diphenyl-2-hydroxypropionic acid[29]

$$Ph \cdot CO \cdot CH_2 \cdot \overset{\displaystyle |}{O} \longrightarrow Ph \cdot CO \cdot \overset{\displaystyle |}{CH} \cdot OH \xrightarrow{-2H} Ph \cdot CO \cdot \overset{\displaystyle |}{CO}$$
$$\quad\quad\quad\quad \overset{|}{C}H_2Ph \quad\quad\quad\quad\quad\quad \overset{|}{C}H_2Ph \quad\quad\quad\quad\quad\quad \overset{|}{C}H_2Ph$$

$$\longrightarrow \quad \underset{Ph \cdot CH_2 \quad\quad CO_2H}{\overset{Ph \quad\quad OH}{\diagdown C \diagup}}$$

Allyl phenacyl ether behaves similarly.

CURTIN effected the rearrangement of desyl ethers[33, 34]

$$Ph \cdot CO \cdot CHPh \cdot OR \xrightarrow[EtOH]{KOH} Ph \cdot CO \cdot CPhR \cdot OH$$

with speed of migration in the descending order,

$$R = \cdot CHPh_2 \quad \Big\rangle \quad \cdot CH_2Ph, \quad -\!\!\!\bigcirc\!\!\!- NO_2 \quad \Big\rangle \quad Ph$$

Rearrangement has also been observed in compounds in which the proton to be extracted is mobilized by adjacent CO_2Et or CN-[29,34]

$$Ph \cdot \overset{\displaystyle |}{CH} \cdot CO_2Et \xrightarrow[EtOH]{NaOEt} Ph \cdot \overset{\displaystyle |}{C}(OH) \cdot CO_2Et$$
$$\quad \overset{|}{O}CHPh_2 \quad\quad\quad\quad\quad\quad \overset{|}{C}HPh_2$$

$$Ph \cdot \overset{\displaystyle |}{CH} \cdot CN \xrightarrow[BuOH]{NaOBu} Ph \cdot \overset{\displaystyle |}{C}(OH) \cdot CN \longrightarrow Ph \cdot CO$$
$$\quad \overset{|}{O}CH_2Ph \quad\quad\quad\quad\quad\quad \overset{|}{C}H_2Ph \quad\quad\quad\quad\quad \overset{|}{C}H_2Ph$$

The rearrangement of benzhydryl ethers[35, 36] presents points of special interest. The initial carbanion can be produced in two exceptional ways:

(a) $Ph_2C(OMe)_2 \xrightarrow{Na} NaOMe + [Ph_2\bar{C} \cdot OMe]\overset{+}{N}a \longrightarrow Ph_2CMe \cdot \overset{-}{O}\overset{+}{N}a$

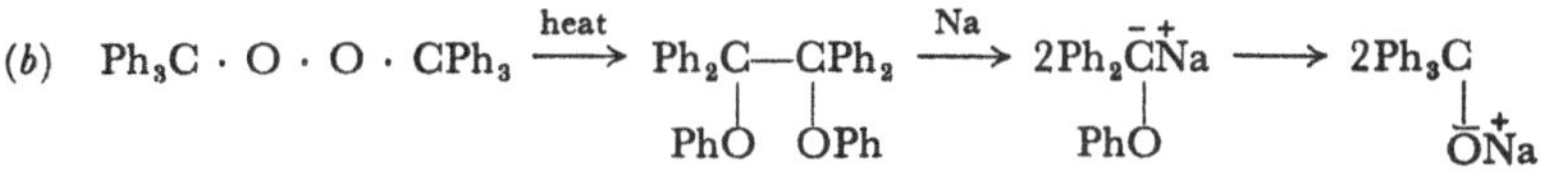

(b) $Ph_3C \cdot O \cdot O \cdot CPh_3 \xrightarrow{heat} \underset{\underset{PhO \quad OPh}{|\quad\quad|}}{Ph_2C—CPh_2} \xrightarrow{Na} 2Ph_2\underset{\underset{PhO}{|}}{\overset{-}{C}}\overset{+}{N}a \longrightarrow 2Ph_3\underset{\underset{\overset{-}{O}\overset{+}{N}a}{|}}{C}$

It is shown that the zinc and magnesium compounds of the type $Ph_2C(OR)$. Metal, colourless and presumably covalent, do not rearrange. In contrast, the alkali metal compounds, coloured and at least in part electrovalent, do rearrange. In dilute solutions the rearrangement proceeds with somewhat diminishing speed as the alkali metal is changed from Li through Na to K; the disparity in favour of Li is much greater in more concentrated solutions. It is suggested that the co-ordination compound $\underset{\underset{R}{|}}{Ph_2C}—O \rightarrow Li$ rearranges specially easily,

and is formed more extensively by the smaller alkali cation. The rearrangement requires a solvent of moderately high dielectric constant (tetrahydrofuran or dioxan); in ether the competing processes (c) and (d) predominate:

(c) $2 Ph_2\bar{C} \cdot OPh \longrightarrow Ph_2C{=}CPh_2 + 2 \bar{O}Ph$

(d) $Ph_2CH \cdot OPh + \bar{R} \longrightarrow Ph_2CH \cdot R + \bar{O}Ph$

With butyl-lithium, benzyl s-butyl ether is rearranged with retention of configuration, but much racemization[35]. On the other hand, benzhydryl 1-phenylethyl ether rearranges with complete retention of configuration

$$\underset{\underset{*CHMePh}{|}}{Ph_2CH \cdot O} \longrightarrow \underset{\underset{*CHMePh}{|}}{Ph_2\bar{C} \cdot O} \longrightarrow \underset{\underset{*CHMePh}{|}}{Ph_2C \cdot OH}$$

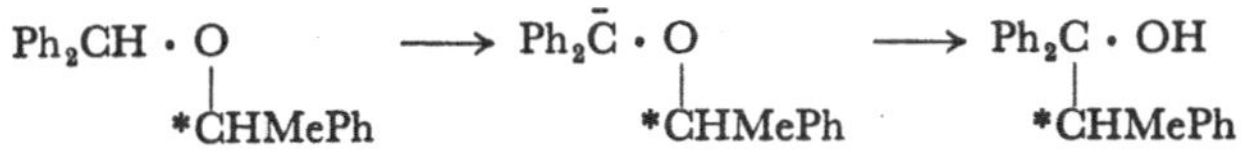

and the successive rearrangements of phenacyl 1-phenylethyl ether proceed with retention of configuration[8]:

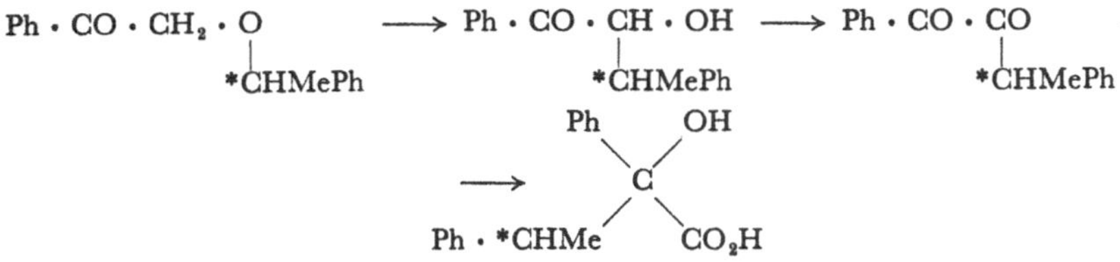

$$\underset{\underset{*CHMePh}{|}}{Ph \cdot CO \cdot CH_2 \cdot O} \longrightarrow \underset{\underset{*CHMePh}{|}}{Ph \cdot CO \cdot CH \cdot OH} \longrightarrow \underset{\underset{*CHMePh}{|}}{Ph \cdot CO \cdot CO}$$

$$\longrightarrow \underset{Ph \cdot *CHMe \quad\quad CO_2H}{\overset{Ph \quad\quad OH}{\diagdown\;\diagup}}{\underset{\diagup\;\diagdown}{C}}$$

Cinnamyl phenacyl ether[8] and cinnamyl fluorenyl ether[29] give on rearrangement the products (XII) and (XIII) respectively, the cinnamyl group migrating without 'inversion'.

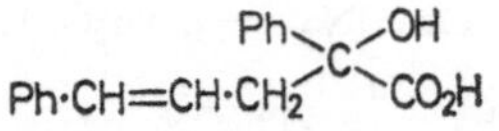

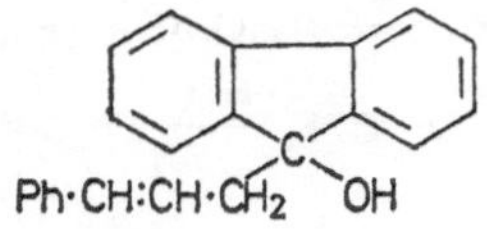

Ph
OH
C
Ph·CH=CH·CH2
CO2H

C
Ph·CH:CH·CH2
OH

(XII) (XIII)

Crotyl fluorenyl ether and 1-methylallyl fluorenyl ether give the same product, 9-crotylfluoren-9-ol [29].

Uncharged Nitrogen to Anionic Carbon

Such rearrangements are much harder to effect than those of ethers[32], partly because the necessary extraction of a proton is more difficult, partly because the change from formally neutral to negatively charged nitrogen is unfavourable in comparison with the corresponding change in oxygen. The latter disadvantage is mitigated when the nitrogen atom bears an aryl group.

The rearrangement of two derivatives of 9-aminofluorene[37] is reported:

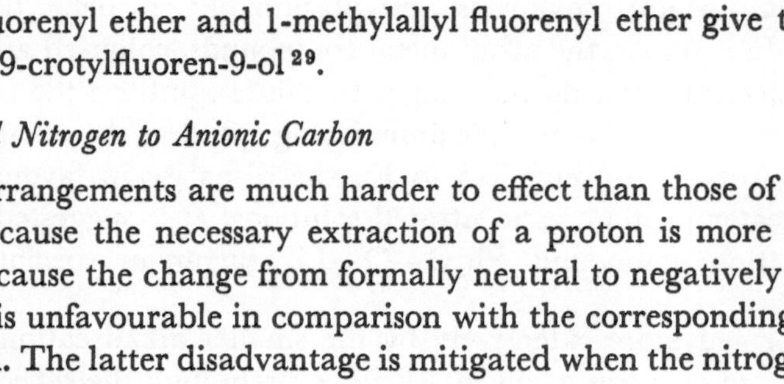

LiAlH4
THF
CH
Me·N·CH2·C10H7
C
Me·NH
CH2·C10H7
[C10H7 =1- or 2-naphthyl]

Phenacylbenzylaniline has been rearranged by mild potash fusion[38]:

$$\mathrm{Ph \cdot CO \cdot CH_2 \cdot N(CH_2Ph)Ph} \xrightarrow[140^\circ]{\mathrm{KOH}} \mathrm{Ph \cdot CO \cdot CH(CH_2Ph) \cdot NHPh}$$

Here the reaction has been shown to be intramolecular by a 'mixing experiment' with isotopically labelled materal

$$\left.\begin{matrix}\mathrm{Ph \cdot CO \cdot CH_2 \cdot NPh \cdot {}^{14}CH_2Ph} \\ \mathrm{Bu^t \cdot CO \cdot CH_2 \cdot NPh \cdot CH_2Ph}\end{matrix}\right\} \longrightarrow \left\{\begin{matrix}\mathrm{Ph \cdot CO \cdot CH(NHPh) \cdot {}^{14}CH_2Ph} \\ \mathrm{Bu^t \cdot CO \cdot CH(NHPh) \cdot CH_2Ph} \quad \text{(XIV)}\end{matrix}\right.$$

The product (XIV) showed no activity. An allyl, but not an alkyl or aryl group will migrate under the same circumstances.

Under very extreme conditions dibenzylaniline and dibenzylmethylamine give respectively 8 per cent and 2 per cent of stilbene, which is known to be produced by similar treatment of the expected product of rearrangement.

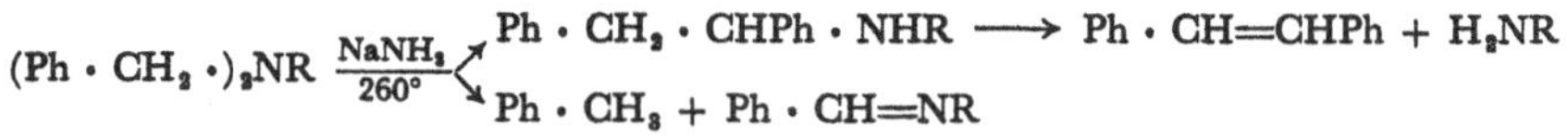

$$(Ph \cdot CH_2 \cdot)_2NR \xrightarrow[260°]{NaNH_2} \begin{cases} Ph \cdot CH_2 \cdot CHPh \cdot NHR \longrightarrow Ph \cdot CH=CHPh + H_2NR \\ Ph \cdot CH_3 + Ph \cdot CH=NR \end{cases}$$

The main process is an elimination reaction leading to a Schiff base.

An application of this reaction gives, by ring-contraction, a tolerable yield of phenanthrene:

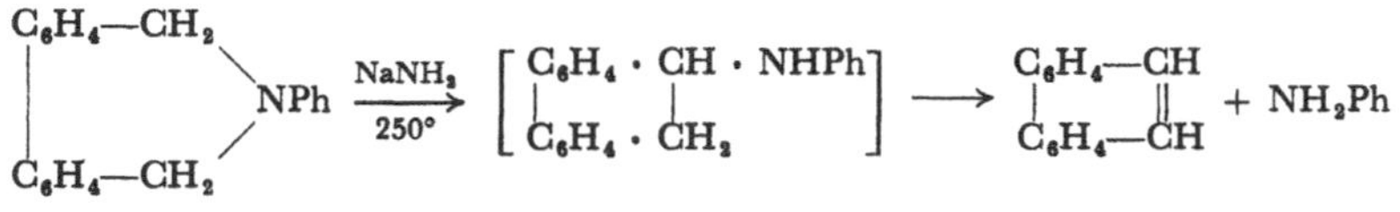

$$\underset{C_6H_4-CH_2}{\overset{C_6H_4-CH_2}{\bigg\rangle}}NPh \xrightarrow[250°]{NaNH_2} \left[\underset{C_6H_4 \cdot \dot{C}H_2}{\overset{C_6H_4 \cdot CH \cdot NHPh}{\bigg|}} \right] \longrightarrow \underset{C_6H_4-\dot{C}H}{\overset{C_6H_4-CH}{\parallel}} + NH_2Ph$$

Some General Considerations

$1 \rightarrow 2$ Rearrangement, and fission by an elimination reaction, appear to be alternative processes:

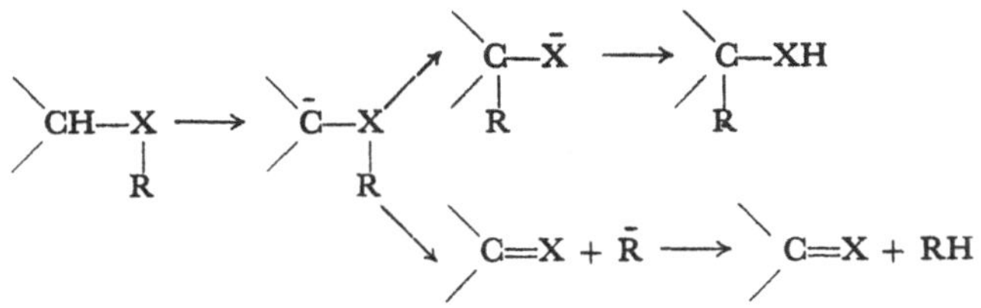

One would expect rearrangement to be favoured by tolerance of X for the unshared electron pair, $\overset{+}{N}R_2 > O > NPh > NMe$, and the elimination by similar tolerance in R. Rearrangement, however, takes place only when on these considerations it is very strongly favoured: basicity of $\bar{X} \ll$ basicity of $\bar{R}$. Compounds of the type $(Ph \cdot CH_2)_2X$ react as follows: $X = \overset{+}{N}Me_2$, rearrangement; $X = O$, rearrangement + elimination; $X = NPh$, elimination + little rearrangement; $X = NMe$, elimination + less rearrangement. It is not surprising that dibenzhydryl ether $(Ph_2CH \cdot)_2O$, undergoes elimination only[39].

It is clear that rearrangement does not proceed by way of preliminary fission, followed by

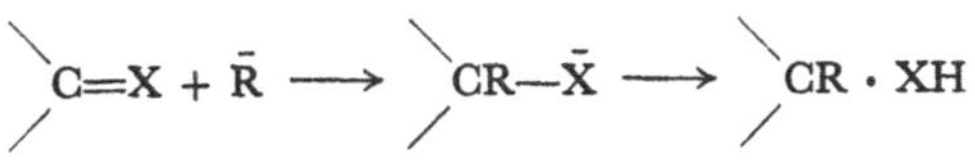

$$\underset{/}{\overset{\backslash}{\diagup}}C=X + \bar{R} \longrightarrow \underset{/}{\overset{\backslash}{\diagup}}CR-\bar{X} \longrightarrow \underset{/}{\overset{\backslash}{\diagup}}CR \cdot XH$$

(a) because rearrangement of the ether can be effected in liquid ammonia which would at once convert $\bar{R}$ into RH, and (b) because rearrangement by R'Li gives at most only minor quantities of $> CR' \cdot OH$.

However, it is quite possible that overall fission by elimination may in some cases proceed by way of rearrangement

$$\begin{array}{ccccccc} \diagdown \\ CH-X & \longrightarrow & \diagdown \bar{C}-X & \longrightarrow & \diagdown C-\bar{X} & \longrightarrow & \diagdown C=X & \longrightarrow & \diagdown C=X \\ \diagup \\ R & & \diagup R & & \diagup R & & \diagup & & \diagup \\ & & & & & & & + \bar{R} & & + RH \end{array}$$

For example, the expected product of the rearrangement of dibenz-hydryl ether, tetraphenylethanol, $Ph_2CH \cdot CPh_2 \cdot OH$, *may* be an intermediate step in the observed fission of the ether to diphenylmethane and benzophenone; the fission products are formed more rapidly from tetraphenylethanol than from the ether[39].

The classification of $1 \rightarrow 2$ electrophilic rearrangements as internal S_N displacements has been energetically opposed [40, 41]. The displacement by frontal attack which this implies is regarded as energetically un-acceptable. While in some cases rearrangement proceeds with sub-stantially complete retention of configuration in an asymmetric migrating radical, in other cases there is extensive racemization. It is suggested that the radical separates as an anion, maintained in a solvent cage, so that it is never kinetically free but may undergo some inversion of configuration before it finally attaches itself to carbon:

$$>\bar{C}-O \longrightarrow \boxed{>C=O} \longrightarrow >C-\bar{O} \longrightarrow >C-OH$$
$$\quad | \qquad\qquad |\bar{R} \qquad\qquad\quad | R \qquad\qquad\quad | R$$

$$>\bar{C}-\overset{+}{N}Me_2 \longrightarrow \boxed{>C=\overset{+}{N}Me_2} \longrightarrow >C-NMe_2$$
$$\quad | R \qquad\qquad\qquad | R \qquad\qquad\qquad | R$$

This formulation is in accord with the very favourable effect of electron-attracting substituents in R.

A very striking confirmation is provided by the behaviour of the ion (XV)[41], in which the asterisked carbon atom is isotopically labelled

$$O={}^*CH \cdot CH_2 \cdot \overset{+}{N}Me_2 \quad \underset{pyridine}{\overset{Hot}{\longrightarrow}} \quad O={}^*CH \cdot \overset{\cdot}{C}H \cdot \overset{+}{N}Me_2 \quad \longrightarrow \quad O={}^*CH \cdot CH=\overset{+}{N}Me_2$$

(XV) $\quad$ CH_2Ph $\qquad\qquad\qquad\qquad\qquad\quad$ CH_2Ph $\qquad\qquad\qquad\qquad$ $\bar{C}H_2Ph$

$$\bar{O}-{}^*CH \cdot CH=\overset{+}{N}Me_2 \qquad\qquad O={}^*CH \cdot \overset{\cdot}{C}H \cdot NMe_2$$
$$CH_2Ph \qquad\qquad\qquad\qquad\qquad CH_2Ph$$
$$\downarrow H_2O \qquad\qquad\qquad\qquad\qquad\qquad \downarrow H_2O$$

43% $\quad$ ${}^*CO \cdot CH_2 \cdot OH \longleftarrow HO \cdot {}^*CH \cdot CHO \qquad\qquad HO \cdot {}^*CH_2 \cdot CO$
$$CH_2Ph \qquad\qquad\qquad\qquad CH_2Ph \qquad\qquad\qquad\qquad CH_2Ph$$
$$\qquad\qquad\qquad\qquad\qquad\qquad\qquad\qquad\qquad 57\%$$

The labelling of the final product indicates simultaneous $1 \to 2$ and $1 \to 3$ migration of benzyl.

WITTIG [42] advocates a somewhat different picture for the migration of an *aryl* radical

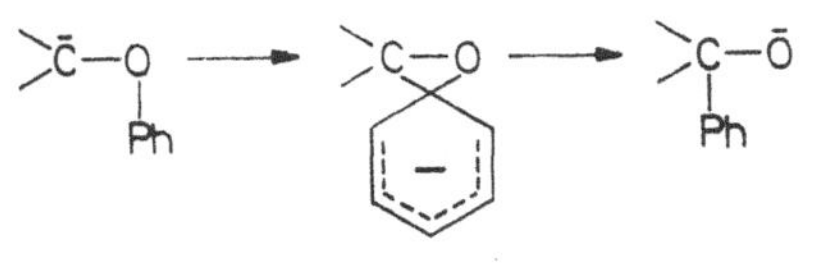

Sulphur to Anionic Carbon

Low-temperature treatment of sulphides, usually with potassamide in liquid ammonia, can cause rearrangement of the Sommelet type[43]:

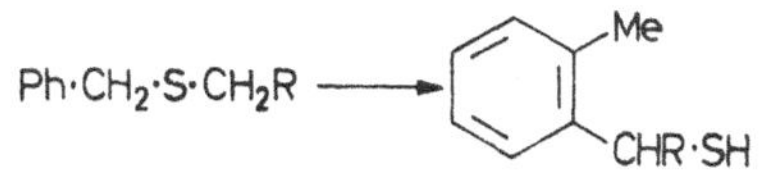

$$Ph \cdot CH_2 \cdot S \cdot CH_2 R \longrightarrow$$

Reaction of sulphides with benzyne (from $o\text{-}C_6H_4F \cdot MgBr$) in tetrahydrofuran[44] leads to $1 \to 2$ rearrangement of sulphonium intermediates:

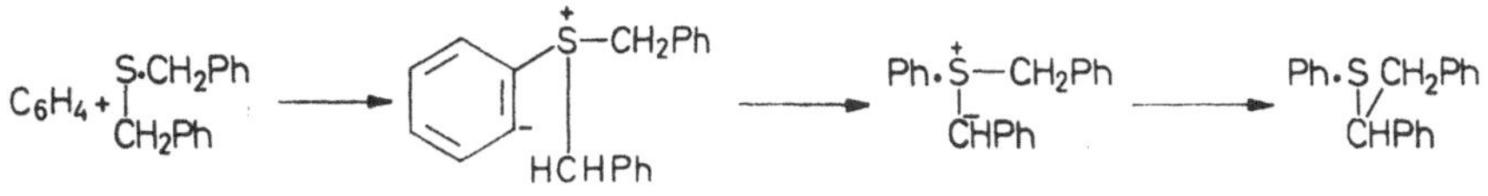

$$R \cdot S \cdot CH_2 \cdot CH{=}CH_2 + C_6H_4 \longrightarrow Ph \cdot \overset{+}{S}R \cdot \overset{-}{C}H \cdot CH{=}CH_2$$

$$[R = Me, CH_2Ph] \qquad\qquad Ph \cdot \overset{+}{S}R \cdot CH{=}CH \cdot \overset{-}{C}H_2$$

$$\longrightarrow Ph \cdot S \cdot CHR \cdot CH{=}CH_2$$

$$\longrightarrow Ph \cdot S \cdot CH{=}CH \cdot CH_2R$$

Note in the last case the alternative $1 \to 4$ migration.

Uncharged Carbon to Anionic Carbon

On general principles, this case should be difficult to realize. In the few recorded examples[45], the carbanion is most powerfully basic, not stabilized by any effective electron-attracting neighbour. Thus,

$$Ph_3C \cdot CH_2Cl \overset{Na}{\longrightarrow} Ph_3C \cdot \overset{-}{C}H_2 \longrightarrow Ph_2\overset{-}{C} \cdot CH_2Ph \longrightarrow Ph_2CH \cdot CH_2Ph$$

61

Metallic 2,2-diphenylpropyls show graded behaviour, ease of rearrangement being correlated with increasing tendency to ionize:

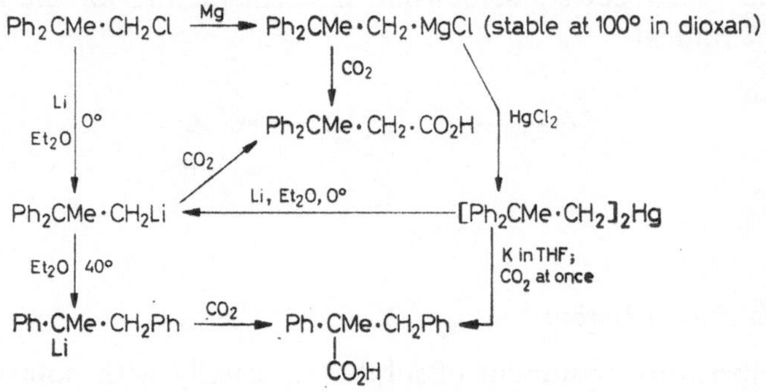

A benzyl group migrates more readily than phenyl:

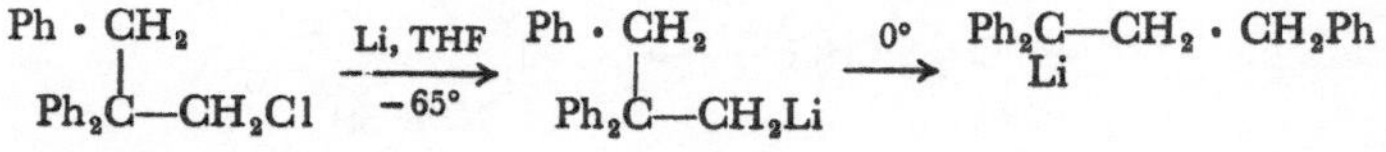

That phenyl migrates faster than methyl in this series (in contrast with, for example, the behaviour of ammonium ions) has been accounted for by molecular-orbital calculations. Qualitatively, in a case like this where the ultimate 'leaving group' —C̄— has little toleration of a negative charge, rearrangement will be favoured if the migrating radical can accommodate a substantial share of such a charge in the transition state—

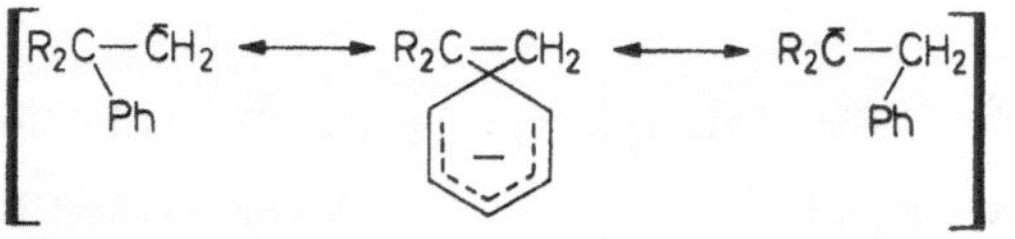

'Onium Sulphur to Anionic Nitrogen

Benzyl and allyl sulphides react with Chloramine T, giving sulphilimines which undergo thermal rearrangement to unstable derivatives of thiohydroxylamine[46]:

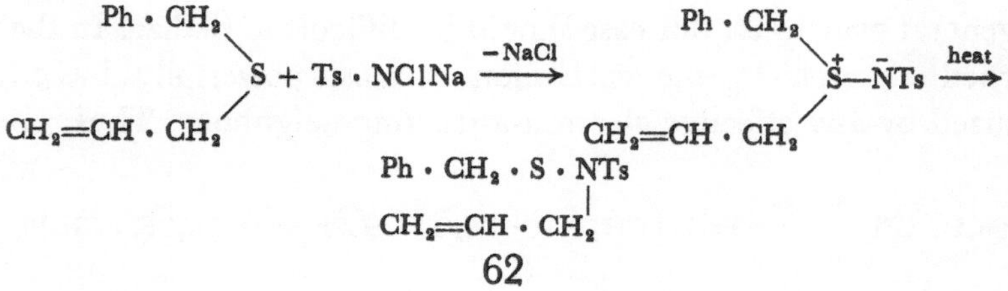

62

Ammonium Nitrogen to Anionic Oxygen

MEISENHEIMER[47] observed that allylmethylaniline oxide and benzyl-methylaniline oxide are rearranged by distillation in steam from alkaline solution:

$$Ph \cdot \overset{+}{N}MeR \longrightarrow Ph \cdot NMe \quad [R = \cdot CH_2Ph, CH_2\text{---}CH\text{=}CH_2]$$
$$\underset{O^-}{|} \qquad \qquad \underset{OR}{|}$$

Saturated alkyl radicals do not migrate under these conditions. The reaction proceeds more easily when, as in Meisenheimer's experiments, an aryl radical is attached to the nitrogen and can conjugate with the lone pair developing on that atom in the transition state. COPE[48] however showed that benzyl- and allyl-dialkylamine oxides rearrange when heated in the anhydrous state. Under gentler conditions, benzhydryl-dialkylamine oxides rearrange very smoothly[49]:

$$Me_2\overset{+}{N}\text{---}CHPh_2 \longrightarrow Me_2N$$
$$\underset{O^-}{|} \qquad \qquad \underset{O\text{---}CHPh_2}{|}$$

All these reactions take place in the absence of excess of alkali. They are shown by 'mixing experiments' to be at least predominantly intramolecular. Substituents in a migrating benzhydryl group affect the speed of rearrangement in much the same way as corresponding substituents in the mobile benzyl group of a rearranging quaternary ammonium ion. When an alkyl radical in an amine oxide contains a β-hydrogen atom, the main thermal reaction is commonly an elimination of Hofmann type, giving an olefin and an NN-dialkylhydroxylamine:

$$\overset{|}{-}CH\text{---}\overset{|}{C}\text{---}\overset{+}{N}R_2 \cdot \bar{O} \longrightarrow \overset{|}{-}C\text{=}\overset{|}{C} + NR_2 \cdot OH$$

An allyl group migrates more rapidly than benzyl, and a radical of this type may undergo 'inversion':

$$Ph \cdot \overset{+}{N}Me \cdot CH_2 \cdot CH \qquad \qquad Ph \cdot NMe \quad CH\text{=}CH_2$$
$$\underset{O^-}{|} \qquad \quad \underset{CHMe}{\|} \longrightarrow \underset{O\text{------}CHMe}{|}$$

Rare cases are known of the rearrangement of nitrones (N-alkylated oximes)[50]:

$$Ph_2C\text{=}\overset{+}{N}\text{---}\bar{O} \longrightarrow Ph_2C\text{=}N\text{---}O\text{---}CHPh_2$$
$$\underset{CHPh_2}{|}$$

63

The reaction is first-order in diethyl carbitol at 200°; energy of activation, 40 kcal.

At least two types of rearrangement, which appear formally to be $1 \rightarrow 2$ electrophilic migrations, are not, in fact, correctly so described. A trialkyl phosphite, when heated with a catalytic quantity of alkyl halide, affords a dialkyl alkylphosphonate, but ARBUSOV[51] showed that the reaction is a two-stage one, involving the alkyl halide:

$$(RO)_3P + R'Br \longrightarrow (RO)_3\overset{+}{P}R'\overset{-}{Br} \longrightarrow (RO)_2PR'{=}O + RBr$$

Many sulphinic esters undergo thermal rearrangement to sulphones, but observations of conditions of reaction and the result of 'mixing experiments' lead to the conclusion that the change proceeds by ionization and recombination[52]:

$$Ar \cdot \overset{O}{\underset{\|}{S}}{-}OCHPh_2 \longrightarrow Ar \cdot \overset{O}{\underset{\|}{S}}{-}\overset{-}{O} + \overset{+}{C}HPh_2 \longrightarrow Ar \cdot \overset{O}{\underset{\underset{CHPh_2}{|}}{\overset{\|}{S}}}\overset{+}{\underset{}{-}}\overset{-}{O}$$

MIGRATIONS OVER LONGER RANGE

Many changes belonging formally to this group are, at least at first sight, simple intramolecular applications of well-known intermolecular processes. Such are alkali-promoted Umesterungen in partially acylated polyhydric alcohols, or the behaviour of o-nitrophenyl benzoate on reduction:

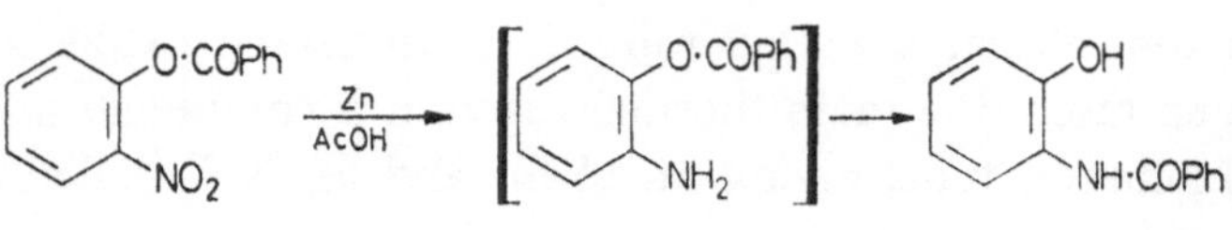

In what follows, attention has been restricted to one or two types of change which have been extensively studied, and to a few isolated examples, interesting from their special connection with formally related $1 \rightarrow 2$ migrations.

The Smiles Rearrangement and Related Changes

A series of rearrangements involving $1 \rightarrow 4$ migration of an aryl group has been studied extensively by SMILES[53, 54]. In the general scheme the group X may be SO_2, SO, S, or O, while Y may be O or NR. The reaction proceeds on mild treatment with alkali. As it is essentially an intramolecular nucleophilic displacement of $\overline{X}$ by $\overline{Y}$, it is greatly favoured by electron-attracting substituents ($2\text{-}NO_2 > 4\text{-}NO_2 >$

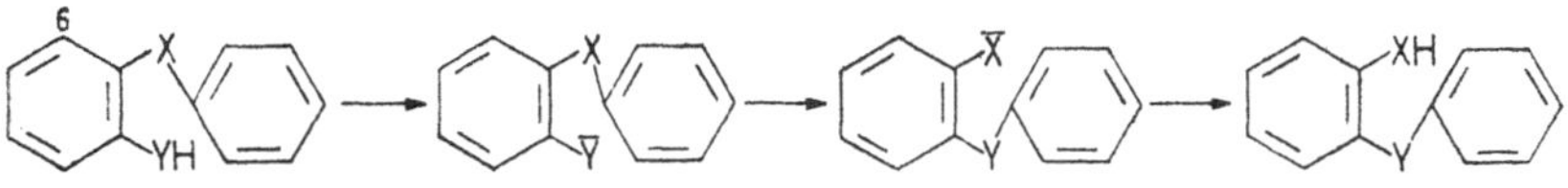

(XVI)

or

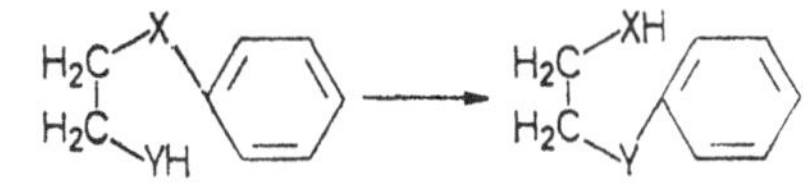

4-SO$_2$Me) in the right-hand benzene ring. The facility of rearrangement falls off as X is changed from SO$_2$ to SO to S, corresponding to the increasing basicity of the resulting anions Ar$\overline{\text{X}}$. The incidence of the rearrangement is summarized in *Table IV*.

Table IV. Incidence of Smiles Rearrangement

YH	X = SO$_2$	SO	S	O	NH
OH	+	—	—	+	
NH$_2$	+	—	—	+	—
NHAcyl	+	+	+		
SH				+	

When Y = O, a proton is immediately extracted by the aqueous alkali, and electron-attracting substituents in the attached ring stabilize the anionic charge and so retard rearrangement; but if Y = NR, ease of proton extraction and reactivity of anion both affect the rate of reaction and are influenced by substitution in opposite senses; in some cases substituents of extreme electron-attracting or donating type are less favourable than those of intermediate character. Bulky substituents in position 6 of formula (XVI) can cause spectacular steric acceleration[54a].

The migrating 'right-hand' nucleus may be replaced by the electron-attracting tetrazolyl system[54b].

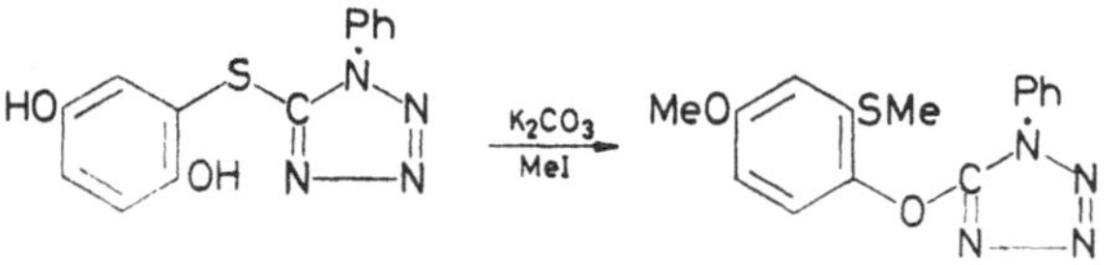

Under conditions of weak to moderate acidity, the rearrangement of a hydroxysulphone may be reversed[55]; possibly:

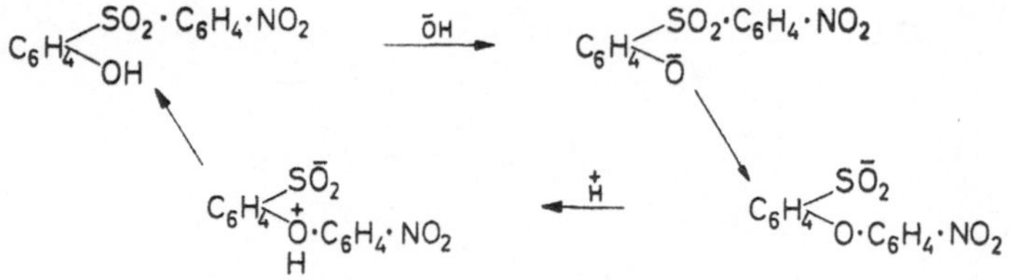

$$C_6H_4 \begin{smallmatrix} SO_2 \cdot C_6H_4 \cdot NO_2 \\ OH \end{smallmatrix} \xrightarrow{\bar{O}H} C_6H_4 \begin{smallmatrix} SO_2 \cdot C_6H_4 \cdot NO_2 \\ \bar{O} \end{smallmatrix}$$

$$C_6H_4 \begin{smallmatrix} S\bar{O}_2 \\ \overset{+}{O} \cdot C_6H_4 \cdot NO_2 \\ H \end{smallmatrix} \xleftarrow{\overset{+}{H}} C_6H_4 \begin{smallmatrix} S\bar{O}_2 \\ O \cdot C_6H_4 \cdot NO_2 \end{smallmatrix}$$

Two successive rearrangements are involved in the cyclic process:

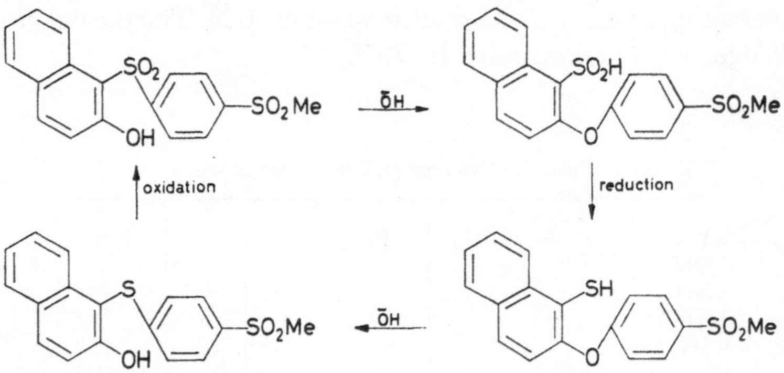

$$\xrightarrow{\bar{O}H}$$

oxidation

reduction

$$\xleftarrow{\bar{O}H}$$

In another case[56], in which X=O, Y=NAcyl, a second migration follows the first:

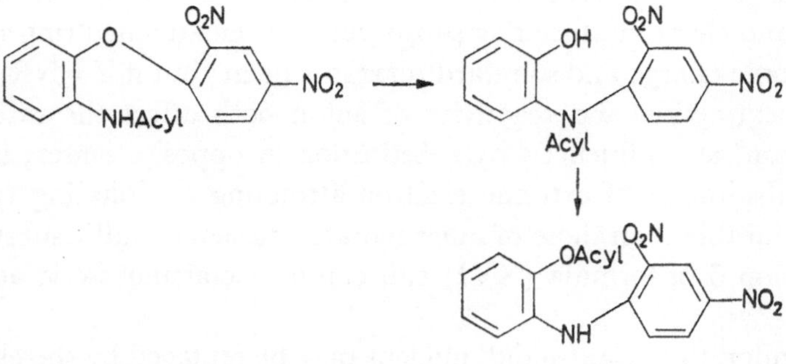

It is not surprising that the first product, an acyldinitrodiphenylamine, can act (intramolecularly) as an acylating agent.

Related 1 → 5 migrations are also recorded[57].

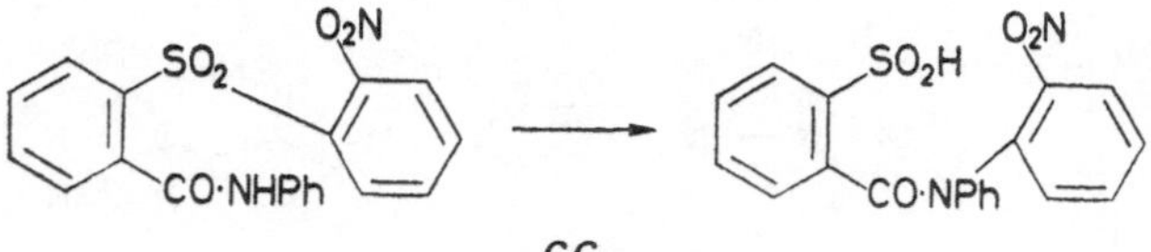

66

The reaction fails in the stereochemically less favoured case in which the left-hand phenylene group is replaced by —CH$_2$—CH$_2$—.

Truce *et al.*[58] has effected similar rearrangements in which the migration terminus is anionic carbon:

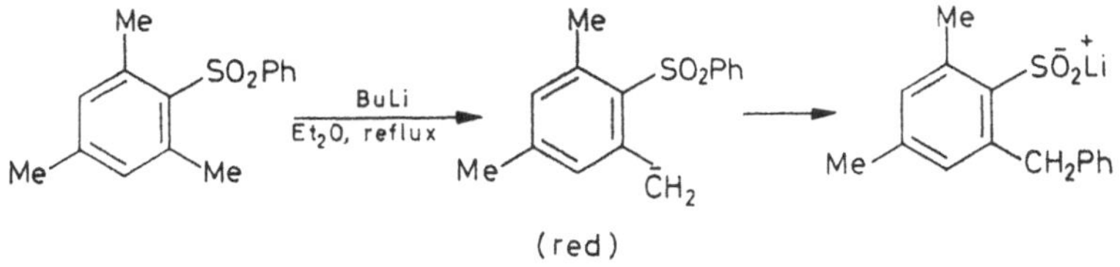

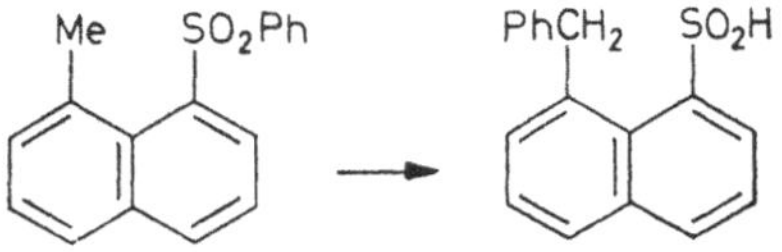

(red)

The reaction shows first order kinetics. Substituents in the migrating nucleus promote rearrangement in the order—*m*-OMe > 3,4,5-Me$_3$ > *p*-Cl > *m*-Cl > *p*-CF$_3$ > *p*-OMe > *m*-CF$_3$. The importance of steric influences is stressed.

Rearrangement of Imino-ethers and Amidines

In these rearrangements[59, 60] which take place at a fairly high temperature, an aryl radical migrates to nitrogen:

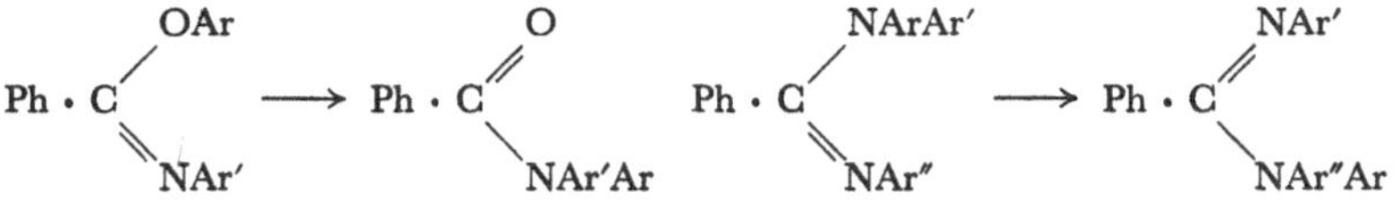

While the migrating radical is transferred to a lone pair on the terminal nitrogen atom, no enduring formal charges are involved, and the changes stand somewhat apart from the typical electrophilic rearrangement so far considered. The reactions show first order kinetics and 'mixing experiments' indicate that they are intramolecular. The amidine change is reversible, while the isomerization of imino-ethers appears to be complete. The reaction may be considered as instigated by attack on the migrating Ar by the lone pair of the recipient nitrogen atom; it is promoted by electron-attracting substituents in the migrating Ar, and hindered by such substituents in the stationary Ar, where they will reduce the basic strength of the recipient nitrogen atom.

An obviously related rearrangement is that of alkyl cyanurates to iso-cyanurates: 1 → 3 migration from oxygen to nitrogen. A more distant

67

relation is the isomerization of thiocyanates $R \cdot S \cdot C{\equiv}N$ to isothiocyanates $S{=}C{=}NR$.

Inversion of an allylic group, and loss of aromatic character of the heterocyclic ring, are involved in the thermal rearrangement of 2-crotyloxyquinoline[61]:

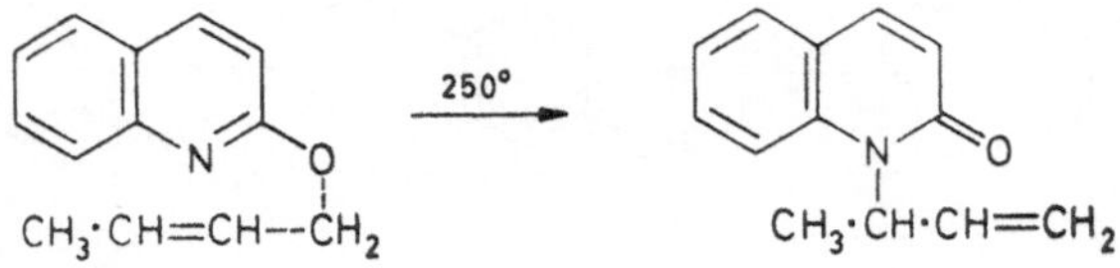

Miscellaneous Rearrangements

The fluorene derivative (XVII) undergoes a $1 \to 4$ migration from ammonium nitrogen to carbon, very easily[62]:

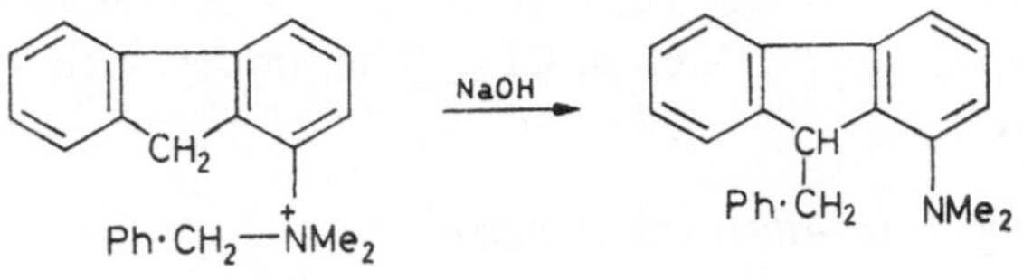

Tertiary amines can react with benzyne[63], giving zwitterions which undergo rearrangement (compare the behaviour of thioethers, *supra*, p. 62):

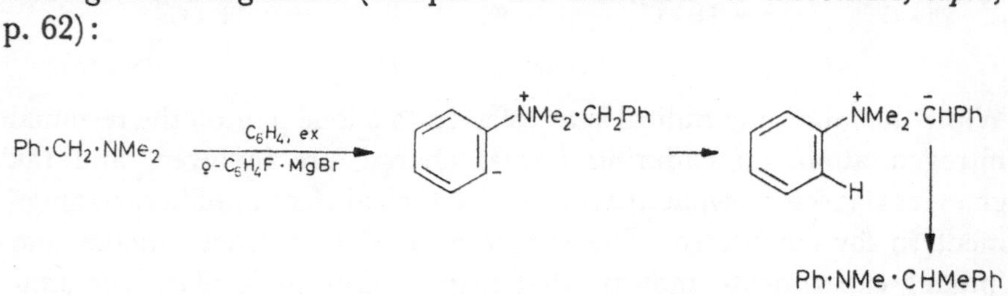

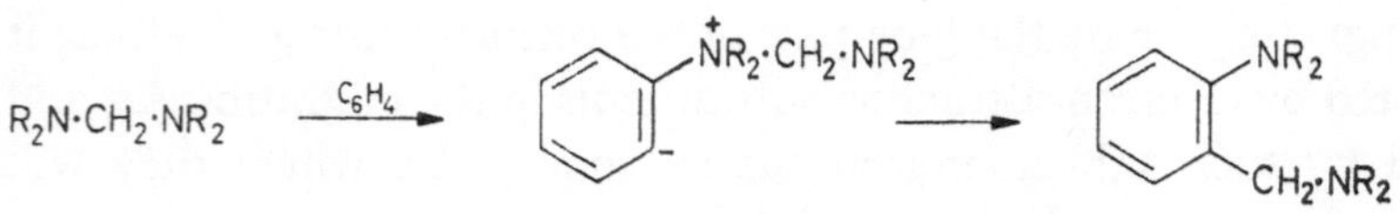

The reaction with allyldiethylamine[64] involves a $1 \to 4$ migration of ethyl:

$$Et_2N \cdot CH_2 \cdot CH=CH_2 \xrightarrow{C_6H_4} \underset{-}{\overset{\overset{+}{N}Et_2 \cdot CH_2 \cdot CH=CH_2}{\bigcirc}} \longrightarrow$$

$$\left. \begin{array}{c} Ph \cdot \overset{+}{N}Et_2 \cdot \bar{C}H \cdot CH=CH_2 \\ \updownarrow \\ Ph \cdot \overset{+}{N}Et_2 \cdot CH=CH \cdot \bar{C}H_2 \end{array} \right\} \longrightarrow \begin{array}{c} Ph \cdot NEt \cdot CHEt \cdot CH=CH_2 \\ \\ Ph \cdot NEt \cdot CH=CH \cdot CH_2Et \end{array}$$

In a similar way, trialkylallylammonium ions show simultaneous $1 \rightarrow 2$ and $1 \rightarrow 4$ migration of an alkyl group in presence of phenyl-lithium[65]; high temperature favours the $1 \rightarrow 4$ change.

BUMGARDNER and FREEMAN[66] formulate the rearrangement of a cyclopropylmethylammonium ion as follows:

$$Ph_2C \overset{CH_2}{\underset{}{\triangle}} CH-CH_2-\overset{+}{N}Me_2 \longrightarrow Ph_2C \overset{CH_2}{\underset{}{\triangle}} CH-\bar{C}H-\overset{+}{N}Me_2 \longrightarrow Ph_2C \overset{CH_2}{\underset{}{\triangle}} CH=CH-\overset{+}{N}Me_2$$
$$\qquad\qquad CH_2Ph \qquad\qquad\qquad CH_2Ph \qquad\qquad\qquad CH_2Ph$$

$$\downarrow$$

$$Ph_2C \overset{CH_2}{\underset{}{\triangle}} CH_2 \cdot CHO \longleftarrow Ph_2C \overset{CH_2}{\underset{}{\triangle}} CH=CH-NMe_2$$
$$\qquad CH_2Ph \qquad\qquad\qquad CH_2Ph$$

Allylvinylammonium ions, produced in a variety of ways, undergo what can be formulated as a $1 \rightarrow 3$ migration of allyl, with inversion[67]:

$$R_2\overset{+}{N} \overset{CH_2 \cdot CH=CH_2}{\underset{CH_2 \cdot CH=CH_2}{\big\langle}} \xrightarrow{\bar{O}H} R_2\overset{+}{N} \overset{CH_2 \cdot CH=CH_2}{\underset{\bar{C}H \cdot CH=CH_2}{\big\langle}} \xrightarrow{H^+} R_2\overset{+}{N} \overset{CH_2 \cdot CH=CH_2}{\underset{CH=CH \cdot CH_3}{\big\langle}}$$

$$\Big\downarrow \bar{O}H$$

$$R_2NH + \overset{CH_2=CH \cdot CH_2}{\underset{O=CH \cdot CH \cdot CH_3}{\big\langle}} \xleftarrow{\text{Hydrolysis}} R_2N \overset{CH_2=CH \cdot CH_2}{\underset{\underset{OH}{CH \cdot CH \cdot CH_3}}{\big\langle}} \longleftarrow R_2\overset{+}{N} \overset{CH_2 \cdot CH=CH_2}{\underset{\underset{OH}{CH-\bar{C}H \cdot CH_3}}{\big\langle}}$$

Finally, some migrations may be cited in which the terminus is anionic oxygen. Thus benzyldimethyl-o-hydroxyphenylammonium ion is rearranged [68] by alkali:

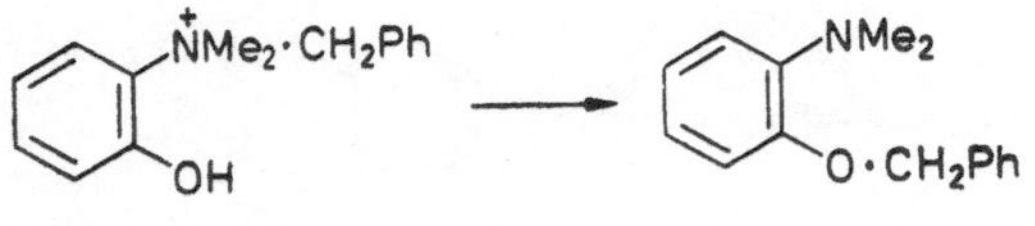

A somewhat similar reaction[69] dismembers a quinuclidine system:

The thermal rearrangement of 2-alkoxypyridine-N-oxides[70] involves the disappearance of formal charges, with concomitant loss of the aromaticity of the pyridine ring:

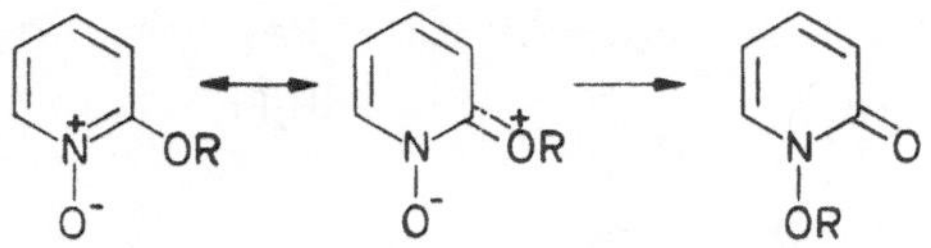

Allyl and benzyl groups migrate at 100°, methyl and ethyl at 140°.

REFERENCES

1 WEITH, W. 'Beziehungen zwischen aromatischen Senfölen und Cyanüren', *Ber. dtsch. chem. Ges.*, 6 (1873) 210–214

2 STEVENS, T. S., CREIGHTON, E. M., GORDON, A. B. and MacNICOL, M. 'Degradation of Quaternary Ammonium Salts. I', *J. chem. Soc.* (1928) 3193–3197

3 THOMSON, T. and STEVENS, T. S. 'Degradation of Quaternary Ammonium Salts. IV', *J. chem. Soc.* (1932) 55–69

4 THOMSON, T. and STEVENS, T. S. 'Degradation of Quaternary Ammonium Salts. VII', *J. chem. Soc.* (1932) 1932–1940

5 DUNN, J. L. and STEVENS, T. S. 'Degradation of Quaternary Ammonium Salts. VI', *J. chem. Soc.* (1932) 1926–1931

6 JOHNSTONE, R. A. W. and STEVENS, T. S. 'Degradation of Quaternary Ammonium Salts. IX', *J. chem. Soc.* (1955) 4487–4488

7 CAMPBELL, A., HOUSTON, A. H. J. and KENYON, J. 'The rearrangement of 1-phenacyl-α-phenylethyldimethylammonium Bromide', *J. chem. Soc.* (1947) 93–95; BREWSTER, J. H. and KLINE, M. W. 'Nucleophilic Displacement via Frontal Attack', *J. Amer. chem. Soc.* 74 (1952) 5179–5182

8 MILLARD, B. J. and STEVENS, T. S. 'Electrophilic Rearrangements', *J. chem. Soc.* (1963) 3397–3403

9 WITTIG, G., TENHAEFF, H., SCHOCH, W. and KOENIG, G. 'Einige Synthesen über Ylide', *Liebigs Ann.* 572 (1951) 1–22

10 HUGHES, E. D. and INGOLD, C. K. 'Influence of Poles and Polar Linkings on the Course Pursued by Elimination Reactions. XIII', *J. chem. Soc.* (1933) 69–75

11 SOMMELET, M. 'Sur un mode particulier de réarrangement intramoléculaire', *C.R. Acad. Sci., Paris*, 205 (1937) 56–58

12 WITTIG, G., MANGOLD, R. and FELLETSCHIN, G. 'Über die Stevens'sche und Sommelet'sche Umlagerung als Ylid-Reaktionen', *Liebigs Ann.* 560 (1948) 116–127

13 KANTOR, S. W. and HAUSER, C. R. 'Rearrangements of Benzyltrimethyl-Ammonium Ion Involving Migration into the Ring', *J. Amer. chem. Soc.* 73 (1951) 4122–4131

14 ZIMMERMANN, H. E. In *Molecular Rearrangements* (P. de Mayo, Ed.) 1st edn, Interscience, New York & London, 1963

15 WITTIG, G. and FELLETSCHIN, G. 'Über Fluorenylide und die Stevenssche Umlagerung', *Liebigs Ann.* 555 (1944) 133–145

16 PUTERBAUGH, W. H. and HAUSER, C. R. 'Isolation of Intermediate Alkali Salt in Rearrangement of Benzyltrimethylammonium Ion', *J. Amer. chem. Soc.* 86 (1964) 1105–1107

17 HAUSER, C. R. and VAN EENAM, D. N. 'Rearrangement to form an *exo*-Methylenecyclohexadieneamine', *J. Amer. chem. Soc.* 79 (1957) 5512–5520, 5520–5524; 'Rearrangements of 2,6-Dimethyl- and 2,3,4,6-Tetramethyl-benzyltrimethylammonium Ions', *J. org. Chem.* 23 (1958) 865–869

18 WITTIG, G. and STREIB, H. 'Zur Erschliessung der Isoindole', *Liebigs Ann.* 584 (1953) 1–22

19 BUMGARDNER, C. L. 'Sommelet–Hauser Rearrangement of Allylbenzyl-dimethylammonium Bromide and Cyclopropylcarbinylbenzyldimethyl-ammonium Bromide', *J. Amer. chem. Soc.* 85 (1963) 73–78

20 LEDNICER, D. and HAUSER, C. R. 'A Novel Ring Enlargement Using Sodium Amide', *J. Amer. chem. Soc.* 79 (1957) 4449–4451; JONES, G. C. and HAUSER, C. R. 'Rearrangement of 1,1-Dimethyl-2-phenylpyrro-lidinium Ion', *J. org. Chem.* 27 (1962) 3572–3576; DANIEL, H. and WEYGAND, F. 'Die Fragmentierung substituierter Pyrrolidiniumver-bindungen mit Phenyl-lithium', *Liebigs Ann.* 671 (1964) 111–118

21 KOHLMEIER, G. and RABINOVITCH, B. S. 'Kinetics of Thermal Isomeriza-tion of *p*-Tolyl Isocyanide', *J. phys. Chem.* 63 (1959) 1793–1794; SCHNEIDER, F. W. and RABINOVITCH, B. S. 'The Thermal Unimolecular Isomerization of Methyl Isocyanide', *J. Amer. chem. Soc.* 84 (1962) 4215–4230; 'The Unimolecular Isomerization of Methyl-d_3-Isocyanide', *J. Amer. chem. Soc.* 85 (1963) 2365–2370; RABINOVITCH, B. S., GILDERSON, P. W. and SCHNEIDER, F. W. 'The Thermal Unimolecular Isomerization of Methyl-d_1-Isocyanide', *J. Amer. chem. Soc.* 87 (1965) 158–160

22 WITTIG, G. and LAIB, H. 'Zur Stevensschen Umlagerung von Onium-salzen', *Liebigs Ann.* 580 (1953) 57–68

23 THOMSON, T. and STEVENS, T. S. 'Molecular Rearrangement in Sulphur Compounds', *J. chem. Soc.* (1932) 69–73; BÖHME, H. and KRAUSE, W. Über Dialkyl-phenacyl-sulfoniumsalze und ihre Spaltung mit wässrigen Laugen', *Chem. Ber.* 82 (1949) 426–432; RUIZ, E. B. 'Reexamination and Mechanism of the Rearrangement of Methylbenzylphenacylsulfonium Hydroxide in weakly alkaline solution', *Chem. Abstr.* 54 (1960) 7623h

24 PINCK, L. A. and HILBERT, G. E. 'Molecular Rearrangement of Fluoryli-dene Dimethyl Sulfide to Fluorene-l-dimethyl Sulfide', *J. Amer. chem. Soc.* 68 (1946) 751–753

25 PASCUAL TERESA, J. DE and SANCHEZ BELLIDO, H. 'Molecular Rearrange-ment of Sulfonium Compounds, II, III', *Chem. Abstr.* 51 (1957) 6537

[26] SCHORIGIN, P. 'Über die Umlagerungen von Benzyläthern, I, II', *Ber. dtsch. chem. Ges.* 57 (1924) 1634–1637; 58 (1925) 2028–2036; 'Weitere Versuche über die Umsetzungen von Äthern mit metallischem Natrium', *Ber. dtsch. chem. Ges.* 59 (1926) 2510–2514

[27] SCHLENK, W. and BERGMANN, E. 'Einige Spaltungsreaktionen mittels Alkalimetallen', *Liebigs Ann.* 464 (1928) 35–42

[28] WITTIG, G., DÖSER, H. and LORENZ, I. 'Über die Isomerisierbarkeit metallierter Fluorenyläther', *Liebigs Ann.* 562 (1949) 192–205

[29] CAST, J., STEVENS, T. S. and HOLMES, J. 'Molecular Rearrangement and Fission of Ethers by Alkaline Reagents', *J. chem. Soc.* (1960) 3521–3527

[30] WITTIG, G. and LÖHMANN, L. 'Über die kationotrope Isomerisation gewisser Benzyläther', *Liebigs Ann.* 550 (1942) 260–268

[31] SCHÖLLKOPF, U. and EISERT, M. 'α-Eliminierung bei alkalimetallorganischen Verbindungen', *Liebigs Ann.* 664 (1963) 76–88

[32] HAUSER, C. R. and KANTOR, S. W. 'Rearrangement of Benzyl Ethers to Carbinols by Potassium Amide', *J. Amer. chem. Soc.* 73 (1951) 1437–1441

[33] CURTIN, D. Y. and LESKOWITZ, S. 'Cleavage and Rearrangement of Ethers with Base. II', *J. Amer. chem. Soc.* 73 (1951) 2633–2636

[34] CURTIN, D. Y. and PROOPS, W. R. 'Cleavage and Rearrangement of Ethers with Base. III', *J. Amer. chem. Soc.* 76 (1954) 494–499

[35] WITTIG, G. and HAPPE, W. 'Über den Einfluss der metallorganischen Bindung auf die Isomerisierbarkeit metallierter Äther', *Liebigs Ann.* 557 (1947) 205–220

[36] WITTIG, G. and STAHNECKER, E. 'Zum Chemismus der Umlagerung von metallierten Benzhydryl-Phenyläthern', *Liebigs Ann.* 605 (1957) 69–93

[37] DAHN, H. and SOLMS, U. 'Über eine Umlagerung tertiärer Amine durch Lithiumaluminiumhydrid', *Helv. chim. acta.* 34 (1951) 907–915

[38] COCKBURN, W. F., JOHNSTONE, R. A. W. and STEVENS, T. S. 'Molecular Rearrangement of Tertiary Amines. I', *J. chem. Soc.* (1960) 3340–3346; JOHNSTONE, R. A. W. and STEVENS, T. S. 'Molecular Rearrangement of Tertiary Amines. II', *J. chem. Soc.* (1960) 3346–3350

[39] CURTIN, D. Y. and LESKOWITZ, S. 'Cleavage and Rearrangement of Ethers with Bases. I', *J. Amer. chem. Soc.* 73 (1951) 2630–2633

[40] SCHÖLLKOPF, U. and FABIAN, W. 'Umlagerungen organischer Anionen. I', *Liebigs Ann.* 642 (1961) 1–21

[41] JENNY, E. F. and DRUEY, J. 'On the Mechanism of the Stevens Rearrangement' *Angew. Chem., Int. edn,* 1 (1962) 155–156; LANSBURY, P. T. and PATTISON, V. A. 'Some Reactions of α-Metalated Ethers', *J. org. Chem.* 27 (1962) 1933–1939; JENNY, E. F. and SCHENKER, K. 'The Mechanism of Anionic Rearrangements—A Stevens-1,3-Rearrangement', *Angew. Chem. (Int. edn.)* 4 (1965) 441–442

[42] WITTIG, G. and CLAUSNIZER, R. 'Zum Chemismus der intraionischen Ätherisomerisation', *Liebigs Ann.* 588 (1954) 145–166

[43] HAUSER, C. R., KANTOR, S. W. and BRASEN, W. R. 'Rearrangement of Benzyl Sulfides to Mercaptans and of Sulfonium Ions to Sulfides Involving the Aromatic Nucleus by Alkali Amides', *J. Amer. chem. Soc.* 75 (1953) 2660–2663

[44] HELLMANN, H. and EBERLE, D. 'Umsetzungen von Thioäthern mit *o*-Fluorophenylmagnesiumbromid', *Liebigs Ann.* 662 (1963) 188–201

[45] GROVENSTEIN, E. 'Preparation of 2-Chloro-1,1,1-triphenylethane and Rearrangement in its Reaction with Sodium', *J. Amer. chem. Soc.* 79 (1957) 4985–4990; GROVENSTEIN, E. and WILLIAMS, L. P. 'Rearrangement in the

Reaction of 2-Chloro-1,1,1-triphenylethane with Lithium and Potassium', *J. Amer. chem. Soc.* 83 (1961) 412–416; 'Rearrangement in the Reaction of 1-Chloro-2,2,3-triphenylpropane with Lithium', *J. Amer. chem. Soc.* 83 (1961) 2537–2541; ZIMMERMANN, H. E. and SMENTOWSKI, A. F. J. 'Carbanion Rearrangements. I', *J. Amer. chem. Soc.* 79 (1957) 5455–5457; ZIMMERMANN, H. E. and ZWEIG, A. 'Carbanion Rearrangements. II', *J. Amer. chem. Soc.* 83 (1961) 1196–1213

[46] ASH, A. S. F., CHALLENGER, F. and GREENWOOD, D. 'The Isomerisation of Diallylsulphilimines. I', *J. chem. Soc.* (1951) 1877–1882; ASH, A. S. F. and CHALLENGER, F. 'The Isomerisation of Diallyesulphilimines. II', *J. chem. Soc.* (1952) 2792–2796

[47] MEISENHEIMER, J. 'Über eine eigenartige Umlagerung des Methyl-allyl-anilin-*N*-oxyds', *Ber. dtsch. chem. Ges.* 52 (1919) 1667–1677; MEISENHEIMER, J., GREESKE, H. and WILLMERSDORF, A. 'Über das Verhalten von Allyl- und Benzyl-amin-oxyden gegen Natronlauge', *Ber. dtsch. chem. Ges.* 55 (1922) 513–522

[48] KLEINSCHMIDT, R. F. and COPE, A. C. 'Rearrangement of Allyl Groups in Amine Oxides', *J. Amer. chem. Soc.* 66 (1944) 1929–1933; COPE, A. C. and TOWLE, P. H. 'Rearrangement of Allyldialkylamine Oxides and Benzyl-dimethylamine Oxide', *J. Amer. chem. Soc.* 71 (1949) 3423–3428; COPE, A. C., FOSTER, T. T. and TOWLE, P. H. 'Thermal Decomposition of Amine Oxides to Olefins and Dialkylhydroxylamines', *J. Amer. chem. Soc.* 71 (1949) 3929–3935

[49] WRAGG, A. H., STEVENS, T. S. and OSTLE, D. M. 'The Rearrangement of Amine Oxides', *J. chem. Soc.* (1958) 4057–4064

[50] COPE, A. C. and HAVEN, A. C. 'Rearrangement of Oxime N-Ethers', *J. Amer. chem. Soc.* 72 (1950) 4896–4903

[51] ARBUSOV, A. E. 'The Conversion of Tervalent into Quinquevalent Derivatives of Phosphorus', *J. chem. Soc.* (*Abstracts*) 92 (1907) 275–276

[52] WRAGG, A. H., McFADYEN, J. S. and STEVENS, T. S. 'The Rearrangement of Sulphinic Esters', *J. chem. Soc.* (1958) 3603–3605

[53] BUNNETT, J. F. and ZAHLER, R. E. 'Nucleophilic Substitution Reactions: The Smiles Rearrangement', *Chem. Rev.* 49 (1951) 362–372; WARREN, L. A. and SMILES, S. 'The Rearrangement of Hydroxy-sulphones. I, II', *J. chem. Soc.* (1931) 2207–2211; (1932) 1040–1047; GALBRAITH, F. and SMILES, S. 'The Rearrangement of Hydroxy-sulphones. V', *J. chem. Soc.* (1935) 1234–1238; EVANS, W. J. and SMILES, S. 'A Rearrangement of *o*-Acetamido-sulphones and -sulphides', *J. chem. Soc.* (1935) 181–188

[54a] BUNNETT, J. F. and OKAMOTO, T. 'Steric Acceleration: The Smiles Rearrangement', *J. Amer. chem. Soc.* 78 (1956) 5363–5367

[54b] CHAPMAN, D. D., JONES, E. T., WILGUS, H. S., NELANDER, D. H. and GATES, J. W. 'The Chemistry of Thioethersubstituted Hydroquinones and Quinones. III', *J. org. Chem.*, 30 (1965) 1520–1523

[55] COATS, R. R. and GIBSON, D. T. 'The Reversibiity of the Rearrangement of *o*-Hydroxy-sulphones', *J. chem. Soc.* (1940) 442–446

[56] ROBERTS, K. C. and DE WORMS, C. G. M. 'A Rearrangement of *o*-Amino-diphenyl Ethers. I', *J. chem. Soc.* (1934) 727–729

[57] EVANS, W. J. and SMILES, S. 'A Rearrangement of Carbamyl-sulphones and -sulphides', *J. chem. Soc.* (1936) 329–331; TOZER, B. T. and SMILES, S. 'A Rearrangement of *o*-Carbamyl Derivatives of Diphenyl Ether', *J. chem. Soc.* (1938) 2052–2056

[58] TRUCE, W. E., RAY, W. J., NORMAN, O. L. and EICKEMEYER, D. B. 'Re-arrangements of Aryl Sulphones. I', *J. Amer. chem. Soc.* 80 (1958) 3625–3629; TRUCE, W. E. and RAY, W. J. 'Rearrangements of Aryl Sulphones, II, III', *J. Amer. chem. Soc.* 81 (1959) 481–484; 484–487; TRUCE, W. E. and GUY, M. M. 'Rearrangements of Aryl Sulphones. IV', *J. org. Chem.* 26 (1961) 4331–4336; TRUCE, W. E. and HAMPTON, D. C. 'Rearrangement of Sulphones, *J. org. Chem.* 28 (1963) 2276–2279

[59] CHAPMAN, A. W. 'Imino-aryl Ethers. III, IV, V', *J. chem. Soc.* 127 (1925) 1992–1998; (1926) 2296–2300; (1927) 1743–1751

[60] CHAPMAN, A. W. 'Dynamic Isomerism Involving Mobile Hydrocarbon Radicals. I–III', *J. chem. Soc.* (1929) 2133–2138; (1930) 2458–2462; 2462–2468; CHAPMAN, A. W. and PERROTT, C. H. 'Dynamic Isomerism Involving Mobile Hydrocarbon Radicals. IV', *J. chem. Soc.* (1932) 1770–1775

[61] MAKISUMI, Y. 'Thermal Rearrangement of Allyl 2-Quinolyl Ethers', *Tetrahedron Letrs.* (1964) 2833–2838

[62] SOER, J. E. and STEVENS, T. S. Unpublished

[63] HELLMANN, H. and UNSELD, W. 'Einwirkung von tert. Aminen auf o-Metallierte Arylhalogenide. I, II', *Liebigs Ann.* 631 (1960) 82–89, 89–94

[64] HELLMANN, H. and SCHEYTT, G. M. 'Einwirkung von tert. Aminen auf o-Metallierte Arylhalogenide. IV', *Liebigs Ann.* 642 (1961) 22–27

[65] HELLMANN, H. and SCHEYTT, G. M. 'Umlagerungen von quartären Ammoniumsalzen', *Liebigs Ann.* 654 (1962) 39–49

[66] BUMGARDNER, C. L. and FREEMAN, J. P. 'Cyclopropylcarbinyl Anions', *Tetrahedron Letrs.* (1964) 737–742

[67] BABAYAN, A. T. and INDJIKYAN, M. H. 'A New Rearrangement-Cleavage Reaction of Quaternary Ammonium Salts', *Tetrahedron.* 20 (1964) 1371–1376

[68] DUNN, J. L. and STEVENS, T. S. 'Degradation of Quaternary Ammonium Salts. VIII', *J. chem. Soc.* (1934) 279–282

[69] GROB, C. A. and RENK, E. 'Untersuchungen in der Chinuclidinreihe, 2. 4-Chinuclidin-carbonsäure', *Helv. chem. Acta.* 37 (1954) 1681–1688

[70] DINAN, F. J. and TIECKELMANN, H. 'Rearrangements of Alkoxypyridine 1-Oxides', *J. org. Chem.* 29 (1964) 1650–1652

4

PHOSPHORYL TRANSFER

V. M. Clark and D. W. Hutchinson

INTRODUCTION	75
PROBLEM	75
RATIONALE	78
CHEMICAL DEVELOPMENT	94
BIOCHEMICAL IMPLICATIONS	100
POSTSCRIPT	111

INTRODUCTION

RECOGNITION of the role played by phosphorus in the nucleic acids and in many coenzymes has led to the development of an organic phosphorus chemistry, concerned mainly with the study of phosphate esters[1] and of the processes of phosphorylation[2]. Only 10 years ago, hardly any mechanistic studies of the reactions of phosphorus compounds had been reported, yet now we have a considerable understanding of many of them[3]: in consequence, interpretations must be based on a detailed knowledge of the bonding in the various types of compound and reaction intermediate. This survey summarizes the present position, and in so doing, develops a chemical classification of phosphoryl transfer reactions, with supporting evidence: for the biochemical sequences appertaining to mitochondrial oxidative phosphorylation, the implications of this classification are considered.

PROBLEM

The transformation of an alcohol or acid, ROH, into the corresponding phosphoryl derivative, $RO \cdot PO_3H_2$ implies the transfer of a phosphorus atom associated with three oxygen atoms, i.e. a phosphoryl group. Many of the energy-conserving processes in living organisms (e.g. oxidative phosphorylation) contain a step in which such a phosphoryl transfer takes place; moreover, the biosyntheses of coenzymes and of nucleic acids and carbohydrates involve analogous stages[4]. In consequence, observations on phosphoryl transfer in both chemical and biochemical systems can be mutually illuminating.

In equation (i) phosphoryl transfer is represented as involving mono-meric metaphosphoric acid, HPO_3.

$$R \cdot OH \xrightarrow{[HPO_3]} RO \cdot PO_3H_2 \qquad \text{(i)}$$

As a discrete chemical entity, this is unknown and, in consequence, reactions leading to phosphorylated products offer a challenge to chemical ingenuity. The same can be said of the transformation of an amine $(R \cdot NH_2)$ into the phosphoramidate $(R \cdot NH \cdot PO_3H_2)$.

Phosphorus, when attached to four highly electronegative ligands, e.g. oxygen, forms bonds with considerable π character[5]. This leads to bond strengthening, *vide infra*. Consequently, the transfer of a (PO_3) group, as opposed to a (PO_4) group, necessitates the cleavage of an apparently strong bond.

In vitro phosphoryl transfer is but one aspect of acylation and, historically, the discoveries of phosphorylating and of carboxylating agents are closely related. The synthesis of various natural products necessitated the development of highly specific acylating agents, and, as peptide synthesis has extended carboxylic acid chemistry, so the synthesis of nucleotides and nucleotide coenzymes has resulted in the development of refined phosphorylating agents[2]. Moreover, since both peptides and nucleotides contain a variety of functional groups develop-ment of methods for their synthesis has been largely dependent on the development of suitable protecting groups[6].

The most striking difference between phosphorylating and carbo-xylating agents lies in the mechanisms of their reactions with nucleo-philes. In the carbon case, the acylating agent $R \cdot CO \cdot \bar{Z}$, (I), is attacked by the nucleophile $\bar{N}^*$ to give the tetrahedral intermediate (II), which can then undergo fragmentation, expelling $\bar{Z}$ with produc-tion of the acylated nucleophile (III).

$$\qquad \text{(I)} \qquad\qquad \text{(II)} \qquad\qquad \text{(III)}$$

On the other hand, the phosphorylating agent $(RO)_2P(O)Z$, (IV), reacts with $\bar{N}$ without intermediate addition, $\bar{Z}$ being displaced in a concerted process.

* Throughout this article $\bar{N}$ will represent a nucleophile in which the superscript implies a lone-pair of electrons. The charge on $\bar{N}$ will not be made explicit in general equations. Likewise, the charge on the leaving group $\bar{Z}$ is implicit, commonly being the same as on $\bar{N}$, when the acylating agent is uncharged.

$$\tilde{N} \quad \begin{array}{c} RO \\ RO \end{array}\!\!> \!\! P \!\! <\!\! \begin{array}{c} O \\ Z \end{array} \quad \longrightarrow \quad \begin{array}{c} RO \\ RO \end{array}\!\!> \!\! P \!\! <\!\! \begin{array}{c} O \\ N \end{array} + Z \ldots \text{(iii)}$$

(IV)

At the phosphoryl centre there is little evidence for the formation of a pentacovalent adduct[7].

There are two important sources of this difference between carbon and phosphorus, namely the possibility (or otherwise) of p_π—p_π bonding, and the use of d-orbitals. Phosphorus forms very few stable compounds with p_π—p_π bonds, whereas the corresponding compounds of carbon, and nitrogen, are legion. Thus, trivalent phosphorus derivatives of types R—P=P—R, R—P=NR, R—P=CR$_2$ and R—P=O, are virtually unknown whereas their carbon, and nitrogen analogues are highly stable.

In the neutral phosphorus atom, the $3d$-orbitals are too diffuse to make an appreciable contribution to bonding but the presence of a formal positive charge on phosphorus can contract them to become commensurate with the $3p$-orbitals[8, 9]. Overlap is then possible.

One of the best examples of this lies in the stability of the phosphinemethylenes (WITTIG reagents)[10], (V).

$$R_3\overset{(+)}{P}\!\!-\!\!\overset{(-)}{C}H_2 \quad \longleftrightarrow \quad R_3P\!\!=\!\!CH_2$$

(V)

Their colours[11] give some indication of the magnitude of the d_π contribution, it being greatest in (VI) and least in (VIII) where the electrons

(VI)	Ph$_3$P=CH$_2$	orange
(VII)	(C$_6$H$_{11}$)$_3$P=CH$_2$	very pale yellow
(VIII)	Ph$_3$P=CH · CO · R	colourless

of the carbanionic moiety are delocalized over the enolate system.

Another line of evidence stems from the quantitative determination of the reactivity of the α-proton in a variety of cations.

Table I. Rates of Deuteration of Tetramethyl Cations of N, P, As, and Sb, and of the Trimethyl Cations of S, Se, and Te[12]

Ion (central atom)	N$^+$	P$^+$	As$^+$	Sb$^+$	S$^+$	Se$^+$	Te$^+$
Exchange rate, log k	$-9{\cdot}75$	$-3{\cdot}37$	$-4{\cdot}61$	$-5{\cdot}62$	$-2{\cdot}44$	$-4{\cdot}09$	$-4{\cdot}63$

Tetramethyl phosphonium and trimethyl sulphonium cations undergo base catalysed deuterium exchange much more rapidly than the tetramethyl ammonium ion.

The orders $P > As > Sb$ and $S > Se > Te$ suggest that the rate is not determined simply by electronegativity differences. The high re-activities of the phosphorus and sulphur-containing cations must be due to conjugation.

Attachment of several electronegative ligands, e.g. oxygen, halogen, can produce an analogous contraction of the d-orbitals. Formation of d_π—p_π bonds is then a relatively favourable process, the d-orbital of the now partially positive phosphorus atom being contracted into the high density region of the p-orbitals of the ligands[8, 9]. This leads, in phos-phoryl derivatives, to a phosphorus-oxygen bond order greater than one. For example, in the case of the mono-sodium salt of phosphor-amidic acid, (IX), crystallographic analysis[13] implies a π-bond order of approximately 0·6 for each of the three phosphorus-oxygen bonds[5].

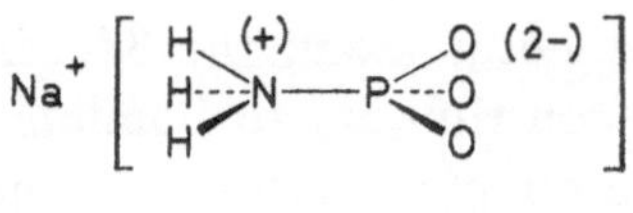

(IX)

For ease of displacement at a phosphorus centre, i.e. for phosphoryl transfer, it would be advantageous to eliminate or attenuate this p_π—d_π contribution. Therein lies the problem.

RATIONALE

Phosphoryl transfer can be achieved when $\overline{Z}$, the group bonded to the phosphorus atom in (IV), is readily displaced as an anion (equation iii) or is eliminated from a reagent of type (X) to give the hypothetical metaphosphate (XI), this then acylating the substrate. Owing to the

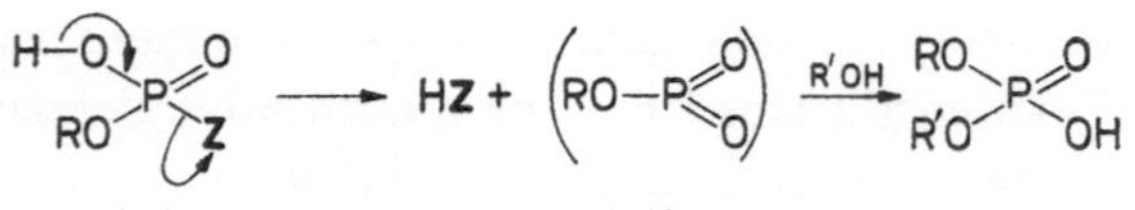

$$\text{(X)} \qquad \text{(XI)} \qquad \ldots \text{(iv)}$$

greater co-ordination number of phosphorus as opposed to carbon, steric hindrance could be more important in phosphorylation than in carboxylation and must be carefully considered before an electronic interpretation of a change in reactivity is adopted. For this reason, most of the published data on the rates of phosphoryl transfer is difficult to analyse. However, the majority of phosphorylation reactions are accepted as being bimolecular, and in this review reaction sequences are so represented. By either route, the leaving group $\overline{Z}$ must accommodate the electron pair originally forming the P—Z bond, and how this can best be done forms the theme of the subsequent discussion.

Reagents of the acid anhydride type must clearly function as phosphorylating agents. Since phosphoric acid is tribasic, a unique bimolecular acylative pathway requires that all hydroxyl groups in the mixed anhydride be protected—otherwise, formation of by-products is unavoidable. In consequence, diesters of phosphorochloridic acid, i.e. $(RO)_2P(O)Cl$, have been widely used with success[2], as have fully esterified pyrophosphates (XII)[14].

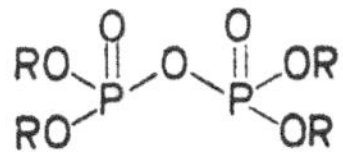

(XII) 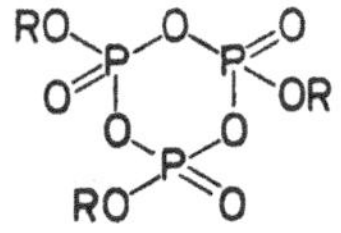   (XIII)

Triesters of trimeta-phosphoric acid (the cyclic anhydride of triphosphoric acid) (XIII) are believed to be the reactive species in phosphorylation reactions carried out with dicyclohexylcarbodiimide[15].

One disadvantage of the mixed acid-anhydride approach to the synthesis of a pyrophosphate lies in the necessity for protection, prior to reaction, of other functions, e.g. alcoholic hydroxyl groups in nucleotides. This has instigated a search for reagents which react specifically with phosphate anions: the most successful of this type so far developed are the monoesters of phosphoramidic acid (XIV)[16, 17], which have

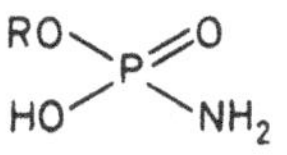

(XIV)

facilitated the synthesis of compounds (e.g. Coenzyme A) containing a large number of potentially reactive functional groups[18].

Phosphoramidates—In phosphoramidates p_π—d_π overlap strengthens the phosphorus–nitrogen bond. Thus, alkaline hydrolysis of diphenyl phosphoramidate (XV; R = Ph) yields the dianion of phosphoramidic acid[19].

Protonation of the nitrogen atom, however, removes this p_π—d_π contribution to bonding and increases the effectiveness of nucleophilic attack at phosphorus: phosphoramidic diesters are converted to phosphorochloridates on treatment with gaseous hydrogen chloride under anhydrous conditions[20]. This behaviour in acid and alkali is in striking contrast to that of carboxylic amides.

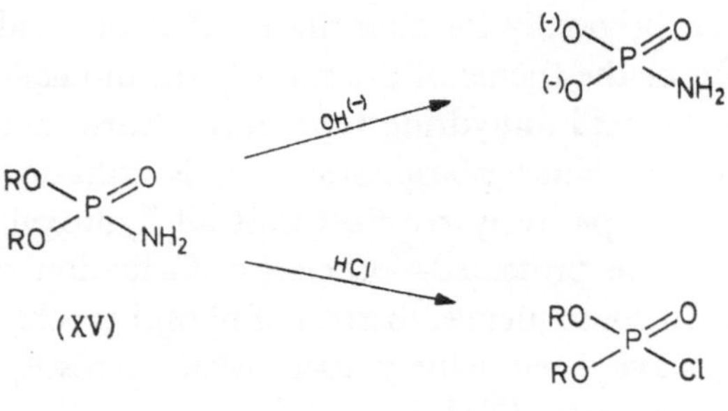

In the case of the monoesters of phosphoramidic acid (XIV) the proton required for P—N cleavage is already present. This can lead, under certain circumstances, to self-phosphorylation[16, 21], but this does not usually interfere with utilization for pyrophosphate synthesis[16, 18]. The general reaction course is as follows:

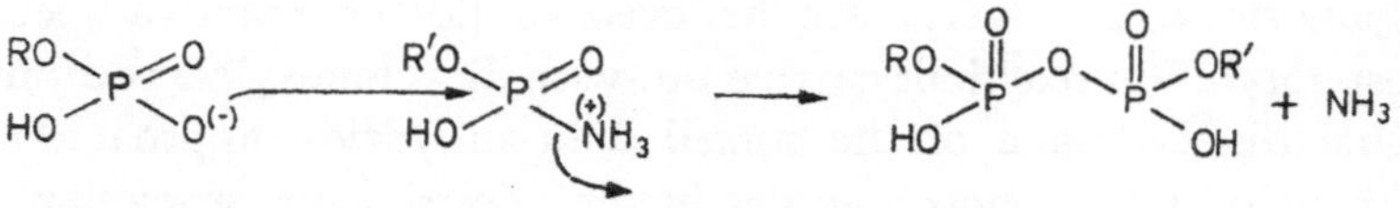

The absence of phosphoryl transfer to solvent when monobenzyl phosphoramidate hemihydrate is heated in benzyl alcohol, with concomitant production of diammonium dibenzyl pyrophosphate in high yield[16], emphasizes the high specificity of this reaction. One is led to ask whether the proton transfer to nitrogen and the attack by oxygen at phosphorus are simultaneous; and whether the transition state in the rate determining step is four-centred[21a]. This would imply retention of

configuration. Monoesters of phosphoroamidic acid contain no intrinsic asymmetry and this thesis cannot be put to the test of experiment. With the sulphur analogue (XVI), P—N cleavage is accompanied by

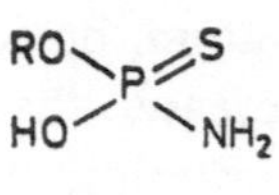

(XVI)

inversion[22]. This, however, does not exclude a different mechanism for the oxygen case since the electronegativity of sulphur is comparable to that of carbon and much less than that of oxygen[23]. Reactions of phosphorus compounds are greatly dependent upon the electronegativities of associated ligands (*vide supra*).

Phosphoroguanidates—The transfer of a phosphoryl group from a phosphoroguanidate also involves the cleavage of a phosphorus to nitrogen bond. With (XIV) in acid solution, this cleavage is facile. However, monoesters of phosphoroguanidic acid do not function as phosphorylating agents under comparable conditions[24]. This marked contrast in reactivity must be attributed to differences in the nature of the P—N bond in the two classes of compounds. With phosphoroguanidates, two prototropic forms (XVII, XVIII) are available. They

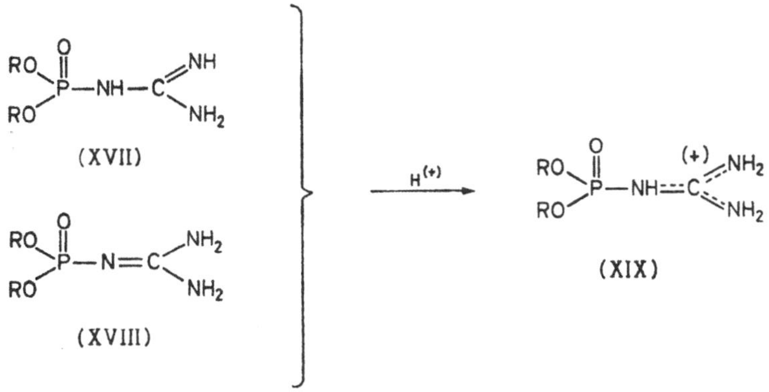

have a common cation (XIX). The formal positive charge can be distributed over the entire guanidinium group and the p_π—d_π overlap in the P—N bond is not decreased to an appreciable extent. If electrons can be withdrawn from the guanidinium moiety by the formation of a complex with a transition metal ion, for example, the p_π—d_π overlap is largely eliminated and metal complexes of phosphoroguanidates can act as phosphorylating agents[25]. This is now an example of a **P—XYZ** system (*vide infra*)[26].

P—XYZ Systems—A molecule of the general formula (XX), where **X, Y** and **Z** are atoms of any element, but commonly hydrogen, carbon, nitrogen, oxygen, sulphur or halogen, is a potential phosphorylating agent if the electrons of the **P—X** bond can be formally accommodated on **Z**. For phosphoryl transfer to take place, the bond order between **X** and **Y** must increase, and that between **Y** and **Z** must decrease, each by one unit.

81

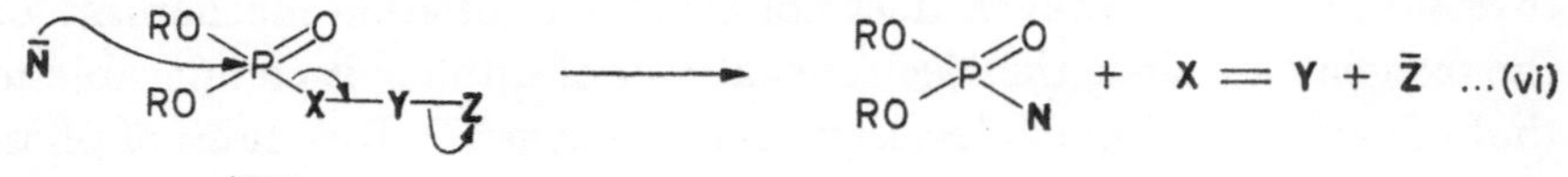

(XX)

$$+ \quad X = Y + \bar{Z} \quad \text{...(vi)}$$

During this reaction, **Z** is the main electron acceptor. This means **Z** must either be strongly electronegative, or must become so in virtue of attack by an electrophile (or an oxidizing agent). Thus, the two essential requirements for a **P—XYZ** system to be a phosphorylating agent are (*a*) that the **P—X** bond should be relatively weak (i.e. that there should be little p_π—d_π bonding between **P** and **X**), and (*b*) that **Z** should attract electrons as strongly as possible.

Condition (*a*) can best be fulfilled if **X** has no lone pair of electrons (for example, $\mathbf{X} = sp^3$ hybridized carbon). If, however, a lone pair of electrons is present on **X**, p_π—d_π bonding between **X** and **P** can be reduced by introducing more effective p_π bonding between **X** and an sp^2 hybridized **Y** atom (for example, in XXI).

(XXI)

Since, in the cases under consideration, the bond order between **X** and **Y** must increase, and that between **Y** and **Z** decrease, vinylogy [i.e. insertion of the group $(\text{—CH}=\text{CH—})_n$] must be confined to the **X/Y** bond: its being placed between **Y/Z** would militate against a reduction in the **Y/Z** bond order. Thus, perphosphoric acid ($\mathbf{X} = \mathbf{Y} = \mathbf{O}$, $\mathbf{Z} = \mathbf{H}$)[27] and the hydroquinone phosphate[28] (XXII) (which can be regarded as a vinylogous perphosphate) become phosphoryl transfer agents under oxidizing conditions.

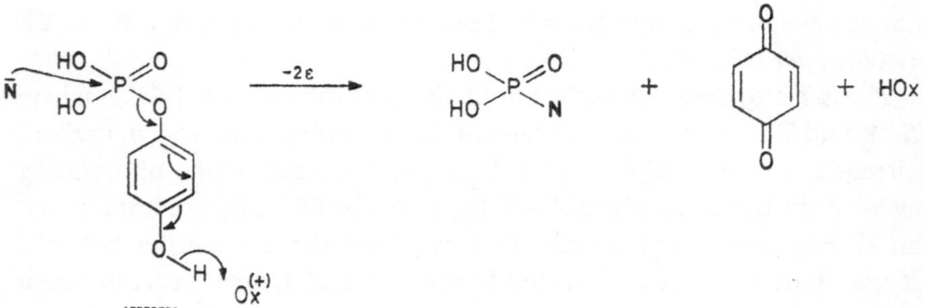

(XXII)

82

An additional advantage of the **P—XYZ** over the **P—Z** system, affecting both the activation energy of the process and the entropy change during the reaction, arises from the extra centres involved in the transition state. For a given **Z** the activation energy can be varied by altering **X** and **Y** without necessarily affecting the overall free-energy change for the reaction.

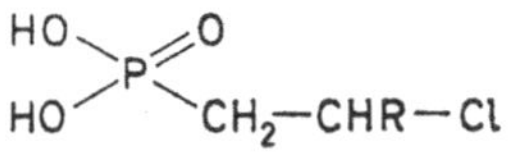

(XXIII) 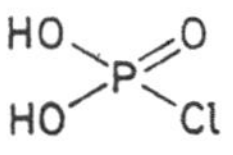   (XXIV)

For example, 2-chlorodecylphosphonic acid (XXIII; **X=Y=C**, **Z**=Cl) is stable in aqueous solutions at pH < 5[28, 29], in contrast to phosphorochloridic acid (XXIV). However, in alkali, or in the presence of organic bases, (XXIII) decomposes rapidly with concomitant phosphoryl transfer[30].

Moreover, if **Y** and **Z** are joined by a single bond, then during phosphoryl transfer by a **P—XYZ** reagent (equation vi) fragmentation will occur and the number of molecules present must increase: the entropy change for this reaction should be positive. Were the **P—XYZ** reagent to undergo unimolecular dissociation, generating monomeric metaphosphate, the entropy change should be even more favourable, and indeed, the (possibly unimolecular) hydrolysis of acetyl phosphate (**X=Z=O, Y=C**) has a low positive entropy change[31].

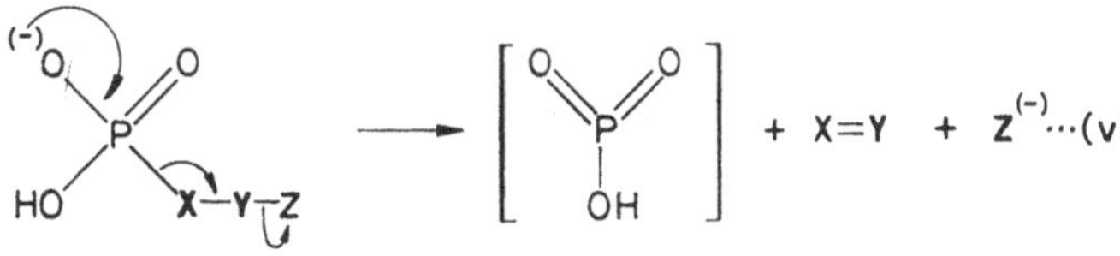

$+ X=Y + Z^{(-)} \cdots$(vii)

Classification of P—XYZ Systems—These systems fall into two main categories, viz., (A) those in which **X** is sp^3—hybridized carbon; and (B) those in which it is sp^2—hybridized carbon, nitrogen, oxygen or sulphur. In (A) there can be no p_π—d_π contribution to the **P—X** bond, whereas in (B) there are various ways in which the attenuation of such a contribution can be made.

CATEGORY A

(a) P—C—C—Hal; β-Halogenoalkylphosphonates

The rapid, alkali-initiated, decomposition of β-halogenoalkylphosphonates has been observed in a variety of cases[32-35] and has been used

83

in phosphoryl transfer to alcohols, phenols and amines[29, 30]. The reaction can even be carried out on β-chloroalk-2-enyl phosphonic acids to give allenes[35]. This reaction is much slower than that observed with

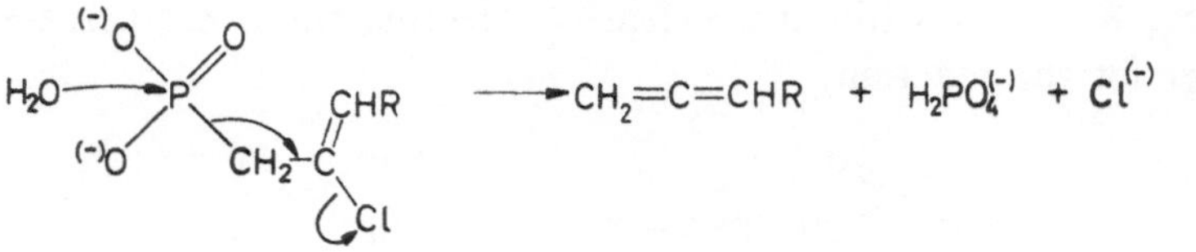

$$H_2O \longrightarrow \cdots \qquad \longrightarrow CH_2{=}C{=}CHR \;+\; H_2PO_4^{(-)} \;+\; Cl^{(-)}$$

the saturated alkyl phosphonates, presumably because of the strengthening of the carbon–chlorine bond by attachment of chlorine to an sp^2- as opposed to an sp^3-hybridized carbon atom.

(b) P—C—C=O; β-Ketoalkylphosphonates

Whereas the phosphorus–carbon bond in saturated alkyl phosphonates is stable to hydrolysis, β-ketoalkylphosphonates yield phosphate and the carbonyl compound under both acid and alkaline conditions[36]. Presumably the ketone is generated as its enol.

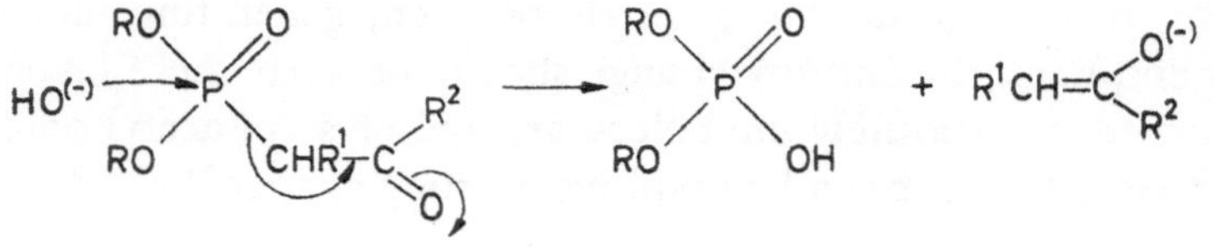

(c) P—C—N=O; p-Nitrobenzylphosphonates

p-Nitrobenzylphosphonic acid decomposes slowly in aqueous alkali to give p-nitrotoluene and trace amounts of 1,2-di(p-nitrophenyl) ethane[37]. The rate of this reaction is presumably reduced by Coulombic repulsion between the anions involved.

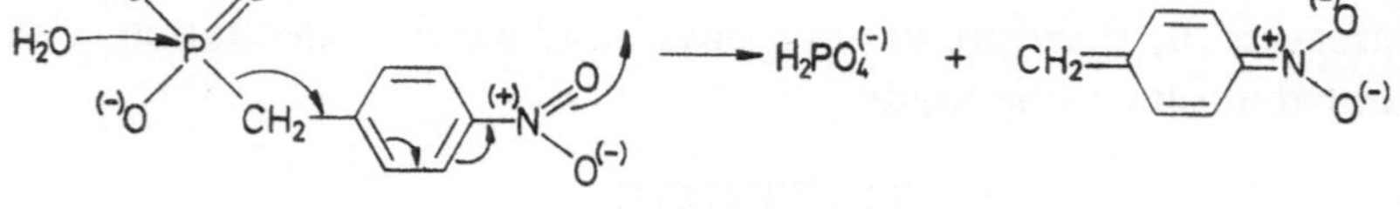

Since in (b) and (c) above, enolic systems are liberated, one might expect the phosphonates of β-diketones to be relatively reactive phosphorylating agents in alkaline solution. This has yet to be shown.

84

CATEGORY B

1. *Attenuation of the p_π—d_π contribution in the Ground State*

(a) *P—N—C=O; N-Acylphosphoramidates*

In contrast to esters of phosphoramidic acid, N-acyl phosphoramidates can phosphorylate alcohols under neutral conditions, in high yield[38]. Acylation of the nitrogen atom reduces the p_π—d_π contribution to the phosphorus–nitrogen bond and, consequently, increases its lability.

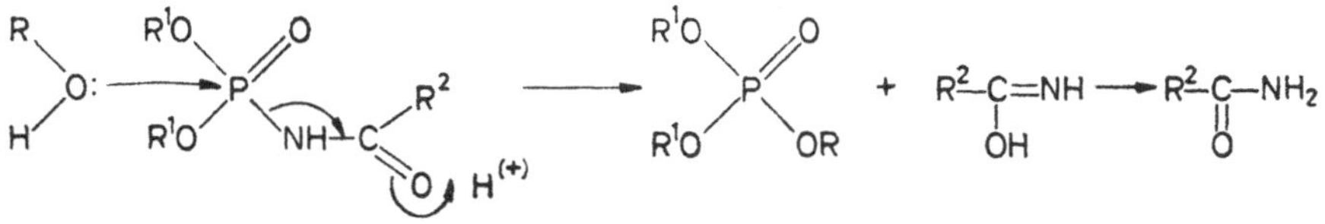

Both N-acylphosphoramidates and N-phosphourethanes[39, 40] undergo hydrolysis in aqueous acid, but here it is difficult to distinguish between a mechanism of the **P—XYZ** type and one involving protonation on nitrogen.

(b) *P—N—N=O; N-Nitrosophosphoramidates*

Nitrosation of N-arylphosphoramidic diesters by either nitrosyl hydrogen sulphate or nitrosyl chloride gives rise to the aryl diazonium phosphate salt[41]. The N-nitroso-N-arylphosphoramidate, which is probably formed initially can react rapidly by an intramolecular process to give the diazonium phosphate ester.

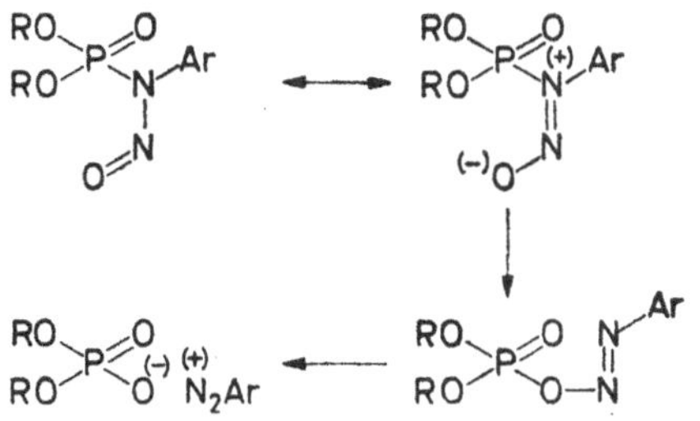

Similarly, salts of monoesters of N-alkyl-N-nitrosophosphoramidic acids produce pyrophosphates on being heated in non-aqueous solvents[42].

(c) *P—O—C=O; Acyl Phosphates*

Mixed anhydrides of carboxylic and phosphoric acids can function either as acylating or phosphorylating agents: reactions of the latter could proceed by a **P—XYZ** mechanism.

85

Nucleophilic attack on an acyl phosphate usually occurs at the carbon atom of the carbonyl group[43] (as this is generally the most electrophilic centre in the molecule) followed by the expulsion of phosphate from the tetrahedral intermediate.

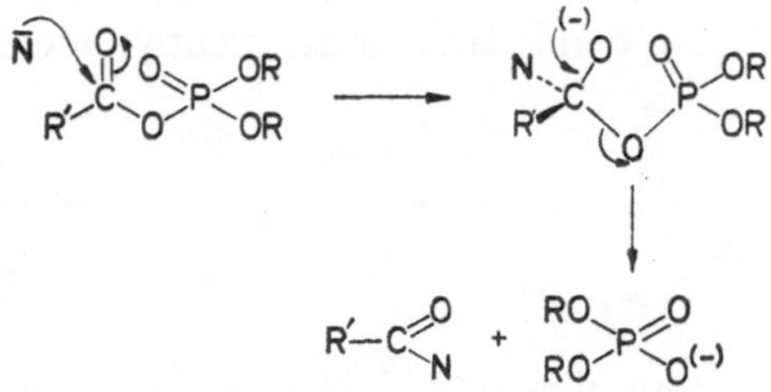

Evidence has, however, been obtained that nucleophilic attack at phosphorus can occur[31], especially during hydrolysis by acid [44-46].

Nucleoside-5'-P^1,P^2-diesters of pyrophosphoric acid have been prepared by the action of acetyl chloride or acetic anhydride on nucleoside-5'-phosphates[47]. The phosphorylating reagent in this reaction must be a mixed anhydride of the nucleoside-5'-phosphoric and acetic acids.

The reaction between isoxazolium salts and carboxylate anions has been used successfully in the synthesis of peptides[48] and the analogous reaction between isoxazolium salts and phosphate anions has been used to effect phosphoryl transfer[49, 50]. The product from the reaction between phosphate and an isoxazolium salt is a vinylogous acyl phosphate.

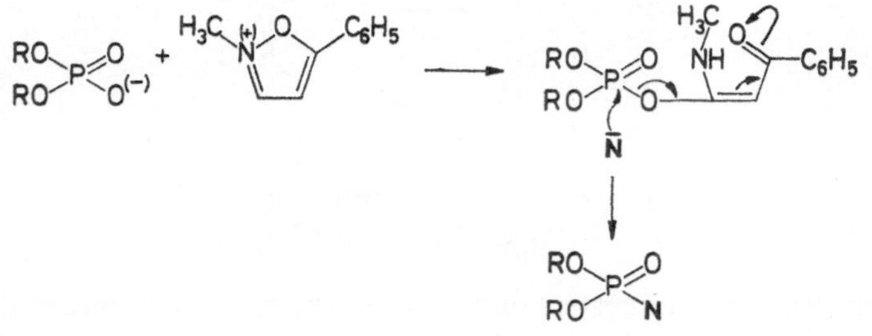

(d) P—O—N=O; Di- and Trinitrophenyl Phosphates

Picryl chloride has been employed in the promotion of phosphorylation reactions by phosphomonoesters[51, 52]. The initial reaction between picryl chloride and a phosphomonoester gives rise to a monoester of picryl phosphate. This can be regarded either as a mixed anhydride of

86

phosphoric and picric acids, or as a vinylogous phosphoronitrate; in the latter case it functions as a **P—XYZ** reagent. Kinetic evidence[53,54] indicates that the hydrolysis of dinitrophenyl phosphates can proceed by a metaphosphate intermediate.

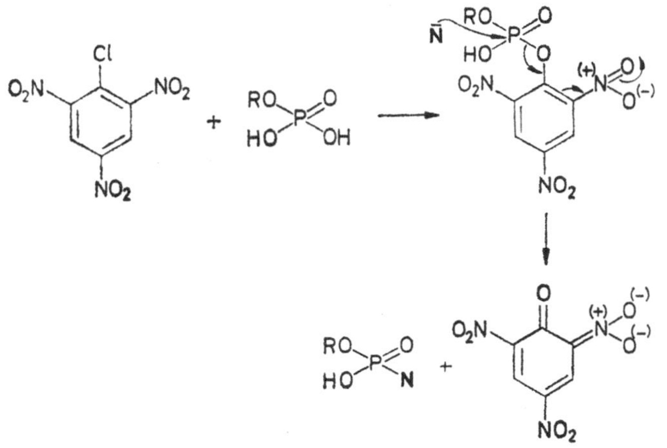

2. *Attenuation of the p_π—d_π contribution by electrophilic attack on* **Z**

(a) *P—N—C=N; Phosphoroguanidates*

The activation of phosphoroguanidates by the formation of complexes with metals has been discussed previously. The formation of a complex with the metal reduces p_π—d_π bonding between phosphorus and nitrogen, enabling phosphoryl transfer to take place[25].

Acylation of the guanidinium moiety has a similar effect on the phosphorus–nitrogen bond. Both N-acetyl and N-benzoyl phosphoroguanidates can function as phosphorylating agents[55].

Moreover, the naturally occurring N-phosphorocreatinine, (**XXV**) an acylated phosphoroguanidate, is also a phosphorylating agent.

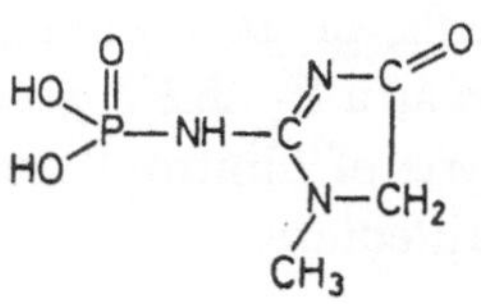

(XXV)

Although N-phosphorimidazoles are formally analogous to N-phos-phoroguanidates as **P—XYZ** systems, they are much more reactive than the latter and are potent phosphorylating agents[56-60]. The presence of a delocalized π system in the imidazole ring must reduce the tendency for p_π—d_π overlap to occur between phosphorus and nitrogen. This should and does increase the reactivity of phosphorimi-dazoles towards nucleophiles.

(b) *P—O—C=C; Enol Phosphates*

Enol phosphates are phosphorylating agents under conditions leading to electron withdrawal from the carbon–carbon double bond[61], e.g. by protonation or oxidation. Hence, an enol phosphate can react with a carboxylic or a phosphoric acid with phosphoryl transfer[62,63]. In

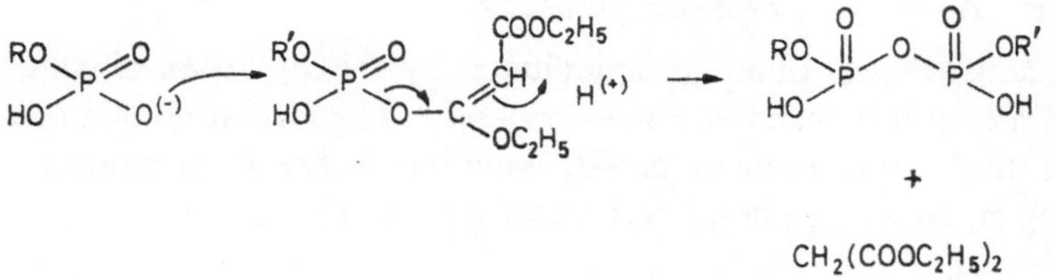

ethanol containing a catalytic amount of mineral acid, a dialkyl ester of an enol phosphate breaks down to the parent carbonyl compound and the ethyl dialkyl phosphate[61]; similarly, aqueous acid leads to the dialkyl phosphate[64].

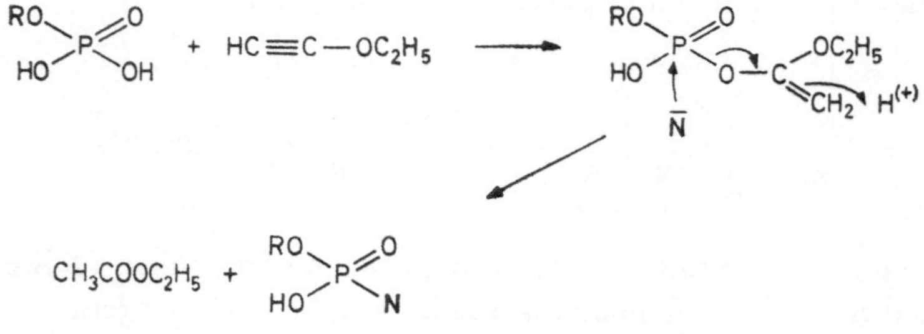

Alkoxyvinyl phosphates, which are intermediates in the reaction between alkoxyacetylenes and monoesters of phosphoric acid, have been used to phosphorylate alcohols[50, 65, 66], and it is believed that the reactive species is an enol phosphate.

(c) *P—O—C=N, Imidoyl Phosphates*

Carbodiimides have been used extensively as reagents for initiating phosphoryl transfer to acids, alcohols, and amines[1]. They function under mild conditions and their hydration products (ureas) are readily removed from reaction mixtures.

The reaction between a carbodiimide and a monoester of phosphoric acid is believed to produce an imidoyl phosphate[15, 67], which is a **P—XYZ** system.

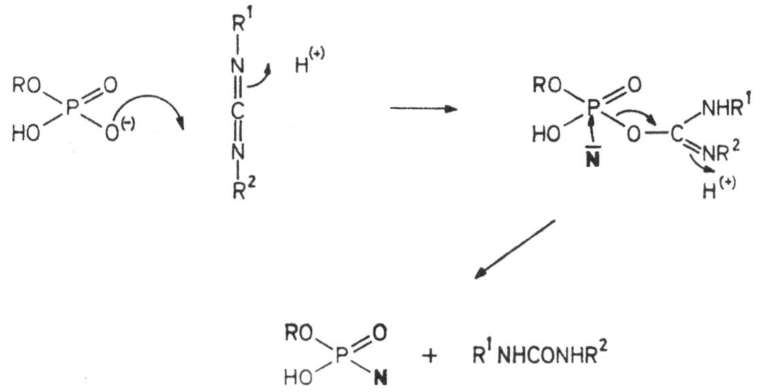

Although imidoyl chlorides are well known[68, 69], the corresponding phosphates have not been isolated and are usually generated *in situ* from a monoester of phosphoric acid and a carbodiimide. A nitrile[70], keteneimine[71], isocyanate[46, 72] or cyanamide[73] can be used in place of the carbodiimide. Also, the Beckmann rearrangement of an aryl-sulphonyl ester of an oxime, in the presence of a diester of phosphoric acid, leads to phosphoryl transfer, presumably via an imidoyl phosphate[74].

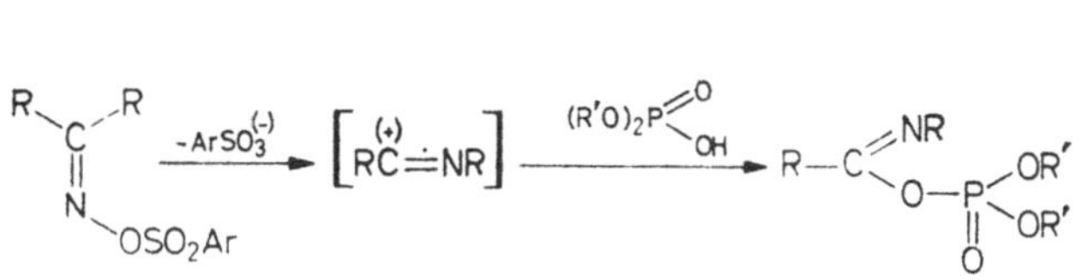

The addition product of phosgene and N,N-dimethyl formamide has been used for the preparation of peptides[75] and in the presence of a monoester of phosphoric acid has been found to be effective in promoting the phosphorylation of alcohols[76]. Here again, an imidoyl phosphate is believed to be a key intermediate.

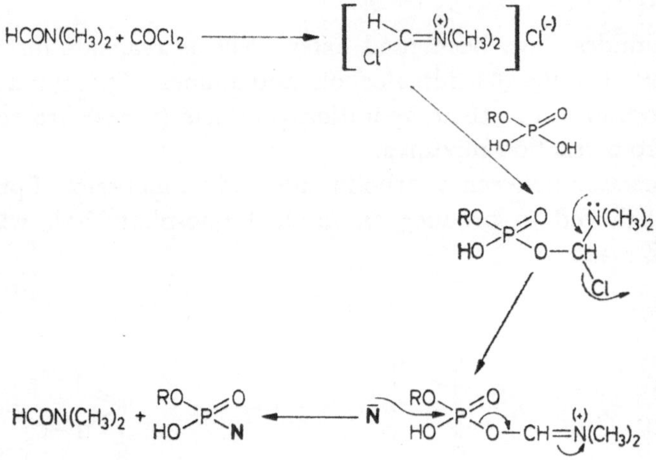

A similar imidoyl phosphate is thought to be involved in the catalysis by N,N-dimethyl formamide of the reaction between an alcohol and a diester of phosphorochloridic acid[77]. α-Pyridyl phosphates, prepared in high yield from a monoester of phosphoric acid and di-α-pyridyl carbonate[78], are phosphorylating agents in the presence of acid[79,80].

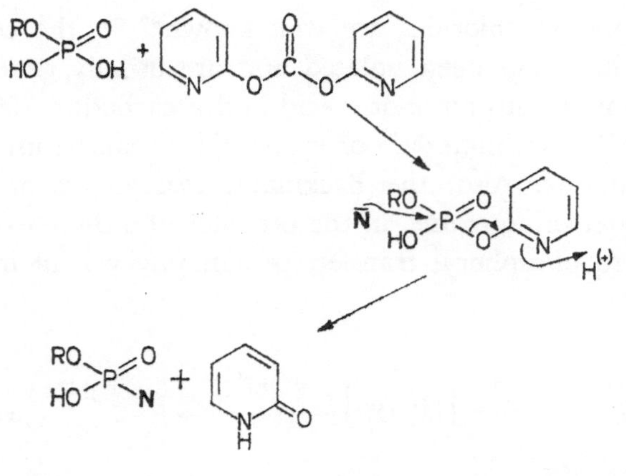

Also, in the presence of cyanuryl chloride, monoesters of phosphoric acid are effective phosphorylating agents[81]; here again, an imidoyl phosphate is a probable intermediate.

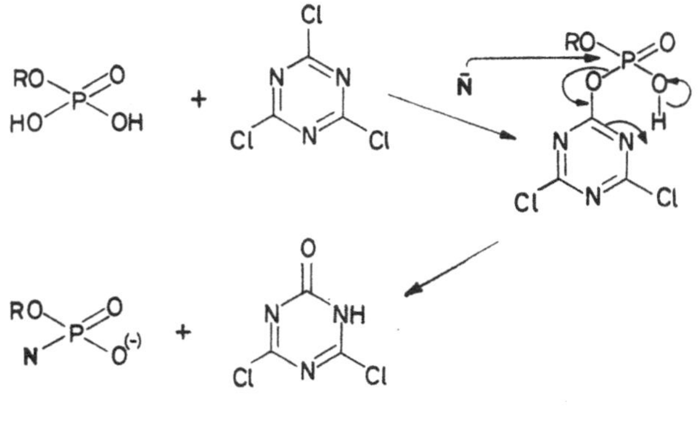

(d) P—O—C—O, 1-Alkoxyalkyl Phosphates

Diethyl phosphoric acid can be converted in high yield into tetraethyl pyrophosphate by the action of ethyl vinyl ether[82]. Presumably, the intermediary 1-ethoxyethyl phosphate breaks down during the phosphorylation step in the following manner:

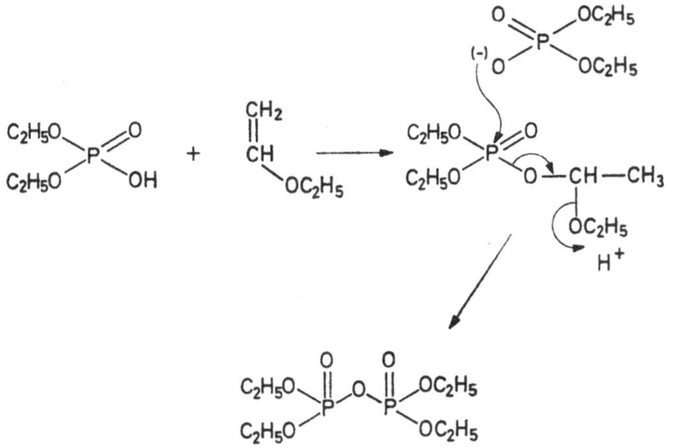

In contrast to hydrolysis in acid, the hydrolysis of α-D-glucose-1 phosphate at pH 6 takes place with almost complete phosphorus–oxygen bond fission[83]. This reaction could take place with concomitant opening of the pyranose ring.

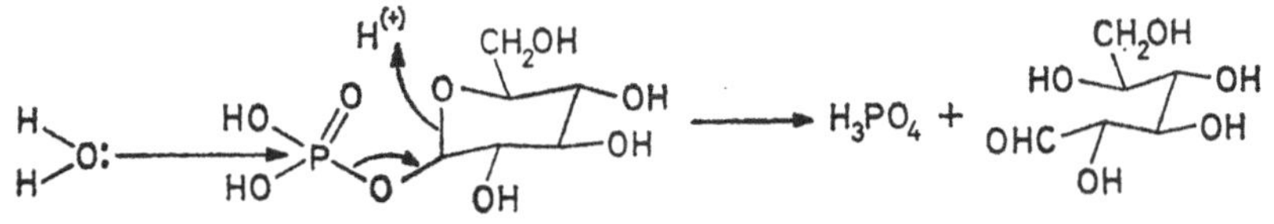

91

3. Attenuation of the p_π—d_π contribution by oxidative attack on Z

(a) P—N—N—H, Phosphorohydrazidates

Phosphorohydrazidates should be phosphorylating agents in acid (i.e. they should behave in a manner analogous to the phosphoramidates), and under oxidizing conditions, when they can be considered as **P—XYZ** systems. Indeed, the oxidation of amino-acid hydrazides has been employed in the synthesis of peptides[84]. Phosphoryl transfer from phosphorohydrazidates has been achieved under both sets of conditions[85]. For example, the oxidation of diphenyl and dibenzyl phosphorohydrazidates with iodine in aqueous pyridine gave quantitative yields of the corresponding phosphoric acid. Two mechanisms (vii) and (viii) are possible. The exact course of the reaction has yet to be determined.

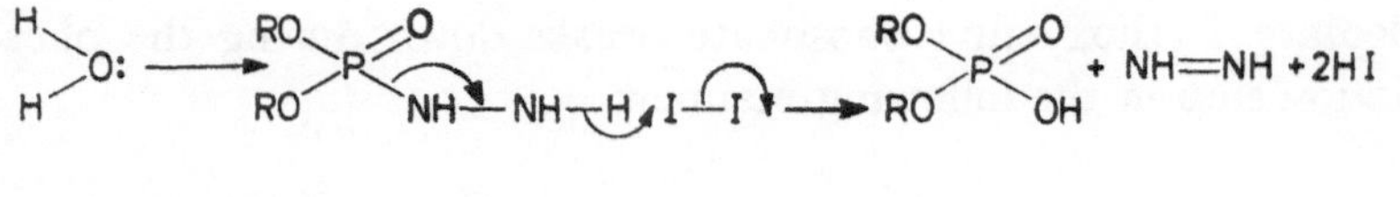

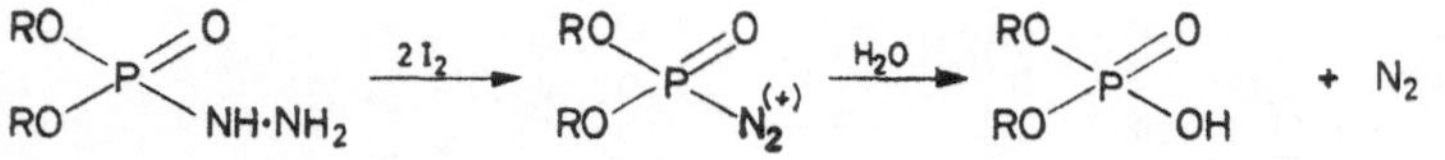

(b) C—O—O—H, Perphosphates and Hydroquinone Phosphates

The reaction between the anticholinesterase Sarin (isopropyl methylphosphonofluoridate) and hydrogen peroxide proceeds by attack of the perhydroxide ion on phosphorus[27]. The peroxyphosphate initially formed is immediately oxidized to isopropyl methylphosphonic acid.

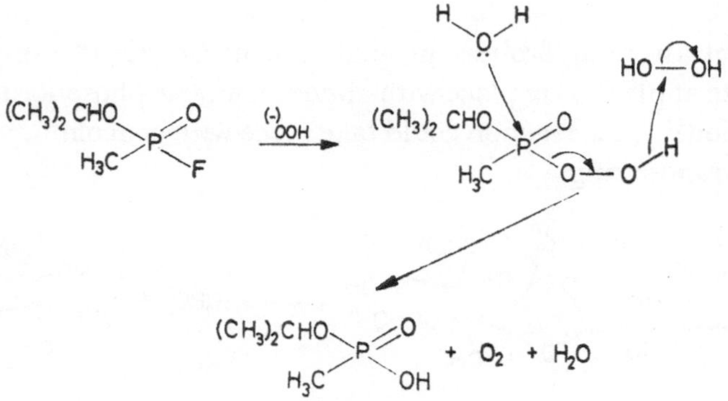

92

Hydroquinone phosphates, as mentioned earlier, are vinylogous perphosphates and transfer phosphate under oxidizing conditions[28,86,87].

The mechanism of the oxidative dephosphorylation is a complex one. Isotopic studies with ^{18}O reveal that approximately 30 per cent phosphorus–oxygen bond fission occurs[88,89]. In addition, the dimethyl ketal of p-benzoquinone has been isolated from the oxidation of hydroquinone monophosphate by ceric ammonium sulphate in methanolic solution[90]. When a dilute solution of durohydroquinone monophosphate in strictly anhydrous methanol is oxidized with an excess of bromine, 10 per cent of monomethyl phosphate is produced, together with 90 per cent of inorganic phosphate[91]. If water is introduced into the system, the yield of monomethyl phosphate rises until a maximum of 40 per cent is reached. This occurs in 55 molar per cent methanol. The yield of monomethyl phosphate then decreases with increasing amounts of water. Similar variations in the yield of monoalkyl phosphates have been observed following the oxidation of dilute solutions of durohydroquinone monophosphate in several different alcohols (ethanol, n- and iso-propanol, n-butanol).

A possible mechanism for this phosphoryl transfer is as follows:

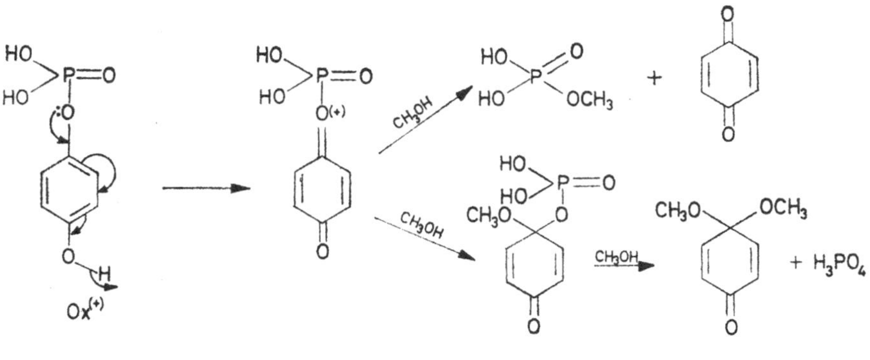

The generation of an oxonium intermediate *prior* to P—O cleavage is in accordance with the observation[28] that bromine oxidation of the acetate-phosphate (XXVI; $R = P(O)(OH)_2$) is virtually instantaneous, giving the bromoquinone (XXVII) whilst the corresponding diacetate (XXVI; $R = CO \cdot CH_3$) is unaffected under comparable conditions. This incorporation of bromine cannot be attributed to the hydrolysis of (XXVI; $R = P(O)(OH)_2$) giving a naphthol since this hydrolysis is known to be slow under the conditions used[92]. Instead, the lone-pair electrons on the phosphorylated oxygen atom must be polarized into the ring before cleavage of the P—O bond can occur. This does not happen in carboxylic esters of phenols, thereby providing yet another instance of the contrast between phosphoric and carboxylic acid derivatives.

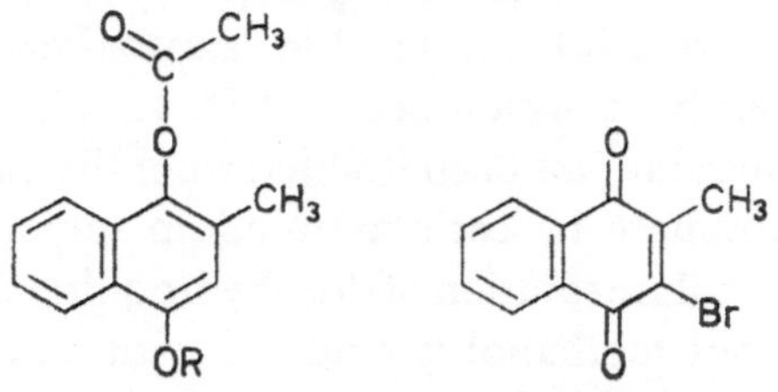

(XXVI)   (XXVII)

Having thus classified the molecular requirements for phosphoryl transfer we propose to examine various chemical developments and bio-chemical implications.

CHEMICAL DEVELOPMENT

Phosphoryl Transfer Promoted by Haloquinones—6,7-Dichloro-5,8-quinoline quinone (XXVIII) promotes acylation by both carboxylic and phos-phoric acids in the presence of transition-metal ions[93]. The correspond-ing anhydrides are produced and esterification of alcohols by carboxylic acids can also be achieved.

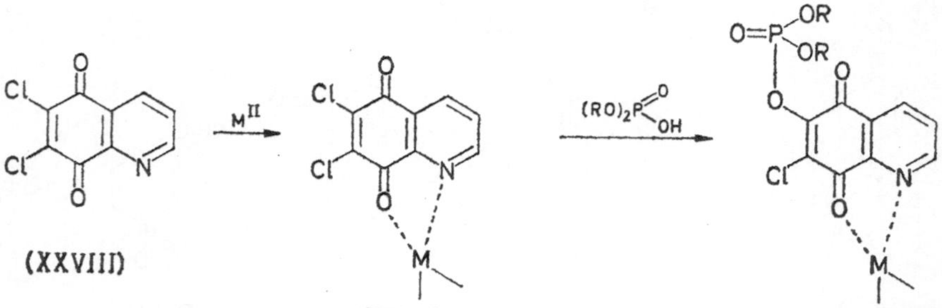

(XXVIII)   (XXIX)   (XXX)

This reaction is cited[93] as an example of acylation catalysed by a complex (XXIX) of the quinone and the transition-metal ion. Through the formation of (XXIX), C_6 of the quinone is made particularly elec-tron deficient, thereby facilitating displacement of chloride by a carboxylate or phosphate anion. Support for this hypothesis was pro-vided by the failure of the quinone (XXVIII) to promote acylation in the absence of a transition-metal ion, and by the instability of the quinone in hydroxylic media in the presence of such ions.

The phosphorylated intermediate (XXX) is, however, a vinylogous acyl phosphate and should be a phosphorylating agent in its own right. If the role of the transition-metal ion is to activate the bond between chlorine and C_6 in (XXVIII), it should be possible to promote comparable acylation using a suitably reactive haloquinone in the *absence* of metal ions. Such has proved to be the case[94]. When several tetramethylammonium salts of orthophosphoric acid are heated in an anhydrous solvent with a high-potential quinone, e.g. 2,3-dichloro-5,6-dicyano-*p*-benzoquinone or tetrachloro-orthoquinone, varying amounts of inorganic pyrophosphate are produced. The yield, depending on the temperature and duration of the reaction, the ratio of quinone to phosphate, and the ionic species of the phosphate, can be as high as 90 per cent[94].

These quinones are also capable of promoting acylation by salts of carboxylic acids; ethyl benzoate is formed, in low yield, when the quinones are shaken in ethanolic solution with tetramethylammonium benzoate. The effect of adding pyridine to the reaction mixture is dramatic[95], the yield of ethyl benzoate rising to 70 per cent. Using the various orthophosphate ions, condensed phosphates are produced, in addition to inorganic pyrophosphate. Complete conversion of phenyl

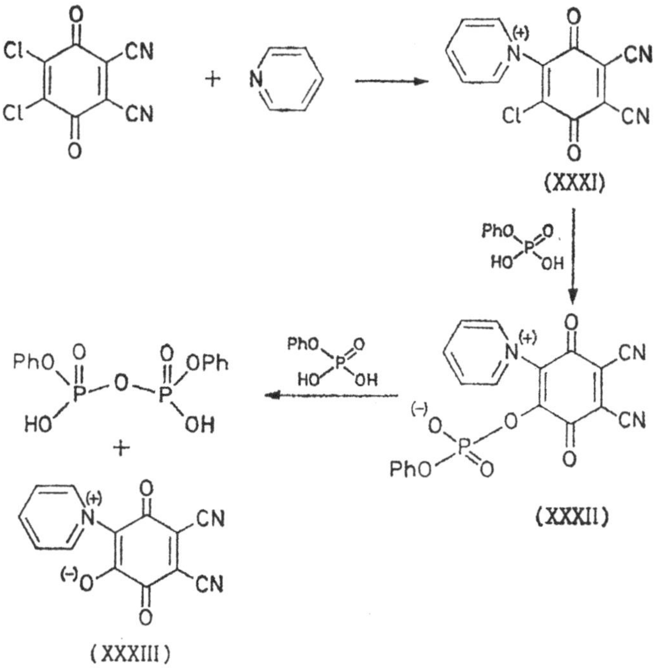

95

dihydrogen phosphate to symmetrical diphenyl pyrophosphate is observed at room temperature. Less reactive quinones (e.g. (XXVIII), chloranil, and 2,3-dichloronaphtho-1,4-quinone) are also effective in promoting phosphoryl transfer. Zwitterionic oxy-pyridinium quinones (e.g. XXXIII) are formed in high yield in this reaction[96]. This suggests that chloride is displaced from the initial adduct (XXXI) by phenyl dihydrogen phosphate to give (XXXII); attack by a second molecule of phenyl dihydrogen phosphate then gives (XXXIII) and the symmetrical diphenyl pyrophosphate.

This pyridine-catalysed reaction is analogous to that involving picryl chloride[52], whereby the pyridinium salt (XXXIV), promotes phosphoryl transfer.

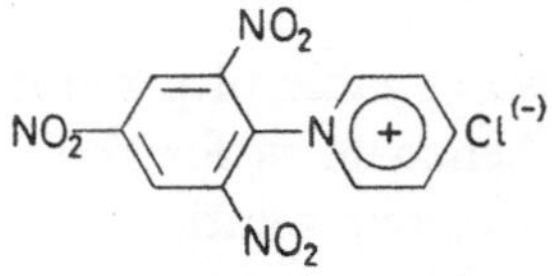

(XXXIV)

With fluoranil, phenyl dihydrogen phosphate in the presence of pyridine gives approximately equal amounts of symmetrical diphenyl pyrophosphate and monophenyl phosphorofluoridate (XXXV)[94]. However, when triethylamine is used as the tertiary base, only the phosphorofluoridate is obtained[94]. The reaction of 2,4-dinitrofluorobenzene with phenyl dihydrogen phosphate in the presence of pyridine or triethylamine gives rise to the same set of products[97]. In this latter

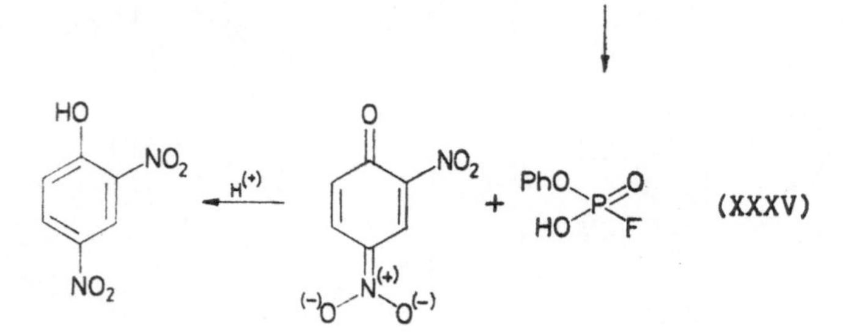

(XXXVI)

(XXXV)

case the formation of the phosphorofluoridate is believed to involve an intramolecular rearrangement of the adduct (XXXVI).

A similar intermediate (XXXVII) should be formed from mono-phenyl phosphate and fluoranil. When either reaction is carried out in pyridine the solvent may also act as a nucleophile. Hence there are two reaction paths available in pyridine.

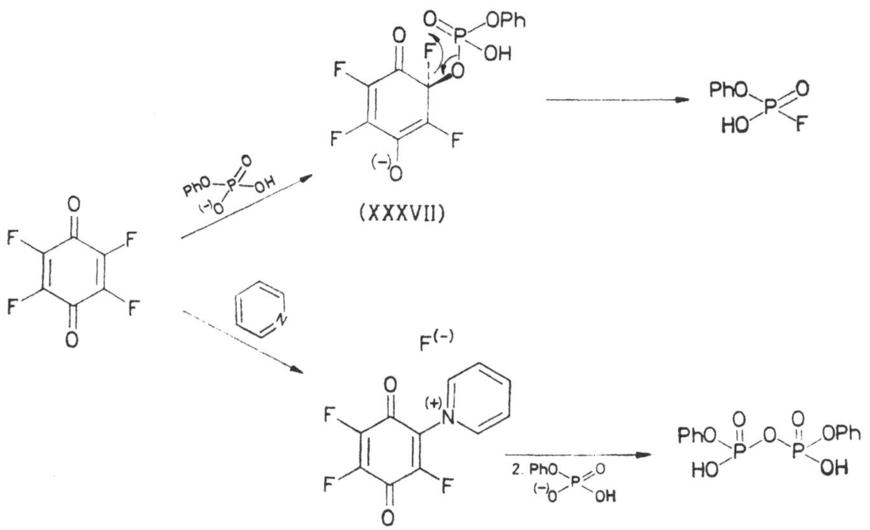

(XXXVII)

Pyrophosphate is *not* produced when triethylamine is added to a solution containing phenyl dihydrogen phosphate and chloranil[95]. During reaction between triethylamine and chloranil *direct* nucleophilic

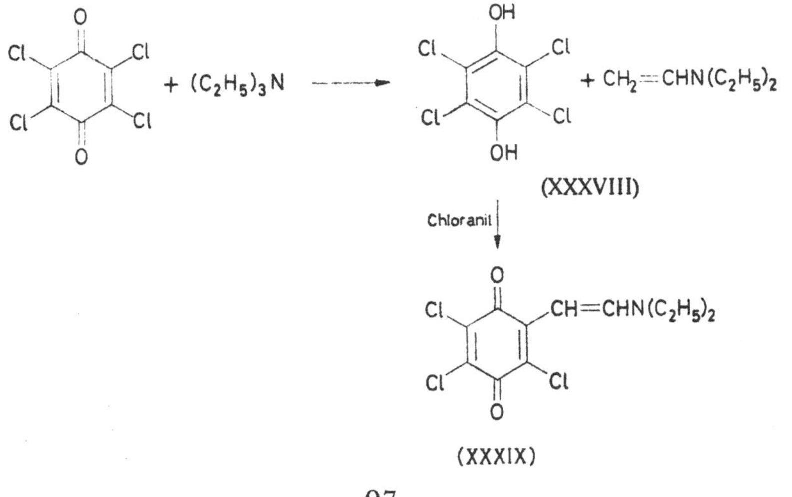

(XXXVIII)

Chloranil

(XXXIX)

97

attack by nitrogen on the quinone ring does not take place. Instead the triethylamine is oxidized by the quinone to an enamine (XXXVIII), which displaces chloride ion from another molecule of chloranil to give the dialkylaminovinylquinone (XXXIX)[98]. Thus, in the presence of triethylamine, the reaction between fluoranil and phenyl phosphate ion will not lead to pyrophosphate since production of analogues of (XXXIX) will supervene.

A methoxyl group can be displaced from a quinone nucleus by hydroxide[99], ethoxide[100] and cyanide ions[101], as well as by amines[99, 102], and cases are known[103] in which an alkoxyl group is replaced in pre-

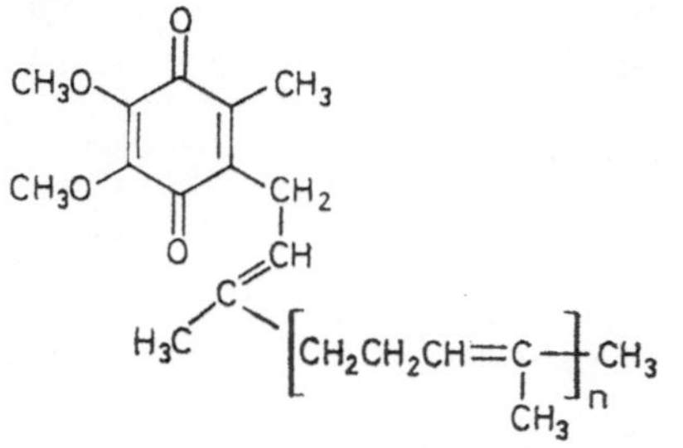

(XL)

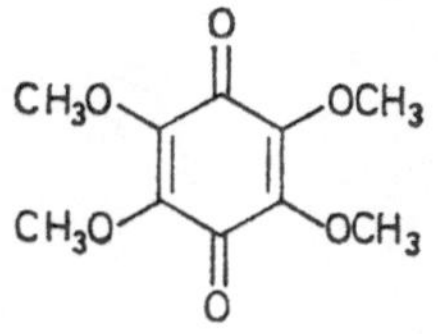

(XLI)

ference to a halogen atom. In view of the occurrence of methoxy-quinones, e.g. the ubiquinones (XL) in mitochondria[104, 105], and their probable participation in oxidative phosphorylation[106], it was of interest to test methoxy-quinones in the general acylation reaction described above. Tetramethoxy-p-benzoquinone (XLI) initiates phosphorylation

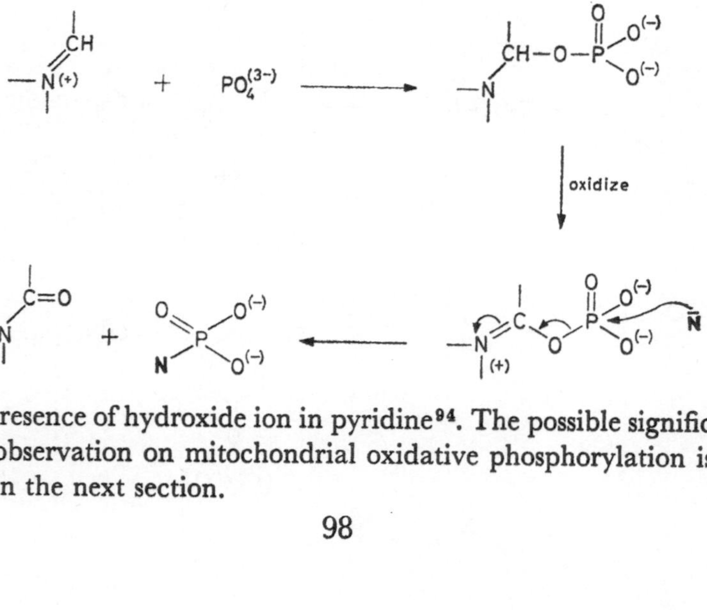

in the presence of hydroxide ion in pyridine[94]. The possible significance of this observation on mitochondrial oxidative phosphorylation is discussed in the next section.

Phosphoryl Transfer Following Addition to $>C{=}\overset{+}{N}{<}$—The trianion of orthophosphoric acid will form adducts with compounds which contain an sp^2-hybridized carbon atom bonded to a planar nitrogen atom bearing a formal positive charge[107]. Under oxidizing conditions they are phosphorylating agents. The probable intermediate is an imidoyl phosphate.

Formation of the adducts, in anhydrous solvents, has been observed with Δ^1-pyrroline-1-oxides (XLII), quinoline-1-oxide (XLIII), and N-benzyl nicotinamide chloride (XLIV).

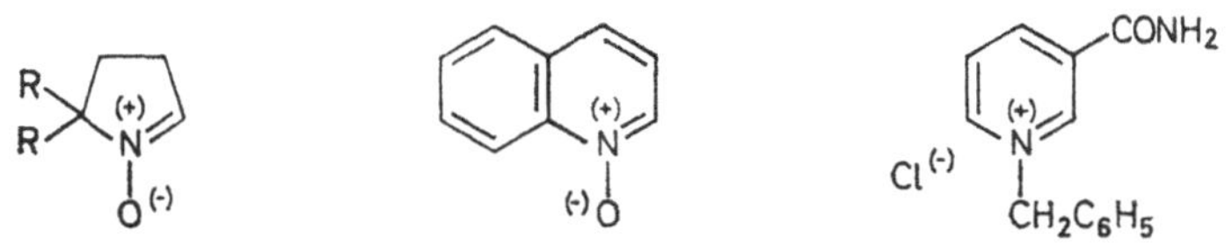

| (XLII) | (XLIII) | (XLIV) |

Although the adducts have yet to be isolated, and indeed, in some cases it appears that their formation is reversed by dilution with anhydrous solvents or water, their oxidation *in situ* results in phosphoryl transfer. The formation of a phosphorylating agent in this manner has important implications in mitochondrial oxidative phosphorylation, an adduct of nicotinamide-adenine dinucleotide (NAD^+) and inorganic phosphate having been observed in mitochondrial nicotinamide-adenine dinucleotide (NAD^+) preparations[108].

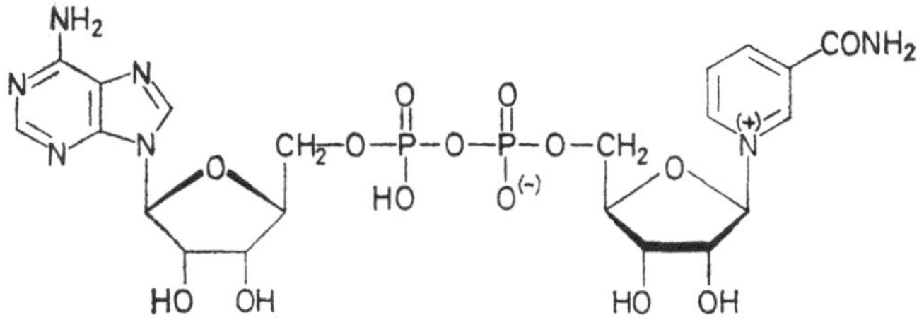

Model experiments with reduced nicotinamide-adenine dinucleotide (NADH) and inorganic phosphate have been carried out in aqueous solution[109]; the yields of inorganic pyrophosphate obtained on oxidation were considerably lower than those described above. It may be pertinent to note that 2-alkyl-4-hydroxyquinoline-1-oxides are inhibitors of mitochondrial oxidative phosphorylation[110, 111], and it is conceivable that the formation of an adduct with inorganic phosphate similar to that described above may be responsible for this inhibition.

99

BIOCHEMICAL IMPLICATIONS

Oxidation of fatty acids, and components of the citric acid cycle, to carbon dioxide and water is one of the major energy-releasing processes in aerobic organisms[112]. In mitochondria, these oxidations are catalysed by the complex assemblage of enzymes of the respiratory chain through which the transfer of a pair of electrons from a substrate (e.g. β-hydroxy-butyrate) to molecular oxygen, results in the formation of three molecules of adenosine triphosphate (ATP) from adenosine diphosphate (ADP) and inorganic phosphate[113-115]. Part of the free energy gained during these oxidative reactions is conserved in the triphosphoryl residue of ATP and is released during the hydrolysis of the latter to ADP and inorganic phosphate.

$$ATP + H_2O \longrightarrow ADP + H_2PO_4^{(-)} \qquad \triangle F^\circ = -7 \text{ kcal}$$

The initial acceptor of electrons from β-hydroxybutyrate is NAD^+ and the second is a flavoprotein whose prosthetic group is flavin-adenine dinucleotide (FAD). Either cytochrome b or ubiquinone is the next acceptor, followed by a series of cytochromes of increasing redox potential until the final transference to molecular oxygen is achieved.

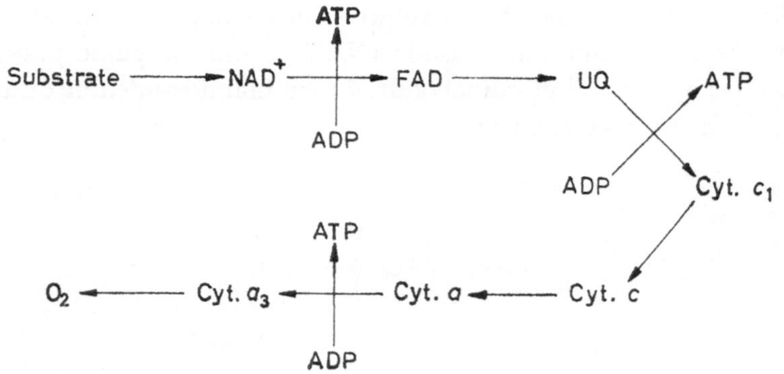

The exact composition of the respiratory chain in the cytochrome b/ubiquinone region is still uncertain, for, whilst there is evidence[116] for the obligatory participation of ubiquinone in the chain, there are marked, unexplained, differences in the kinetics of oxidation and reduction of ubiquinone in phosphorylating and non-phosphorylating systems[117]. Cytochrome b and ubiquinone may represent alternative pathways for the passage of electrons from the flavoprotein to cytochrome c_1[118].

Phosphoryl Transfer at the NAD⁺/FAD Site of ATP Synthesis—Evidence has been presented[119] for the existence in mitochondria of a chemical species involving nicotinamide-adenine dinucleotide which does not react with alcohol dehydrogenase but which can be converted into NAD⁺ after incubation of the mitochondria with inorganic phosphate, ADP, or 2,4-dinitrophenol (a potent inhibitor of oxidative phosphorylation). In addition, a so-called 'high-energy' form of NAD⁺ has been described which, on incubation of mitochondria with inorganic phosphate and ADP, is capable of promoting the synthesis of ATP[120].

The presence of a phosphorylated derivative of NADH (the reduced form of NAD⁺) and its participation in oxidative phosphorylation has been demonstrated[108] by the incubation of mitochondria with ³²P-labelled inorganic phosphate. Small amounts of a highly unstable, ³²P-labelled, derivative of NADH could be detected and the labelled phosphoryl group could be transferred enzymically to ADP. From spectroscopic evidence[108] and tritium labelling experiments[121], the phosphate residue is presumed to be attached to the 6-position of the nicotinamide ring. This addition of phosphate and subsequent phosphoryl transfer under oxidizing conditions has been represented as follows[109].

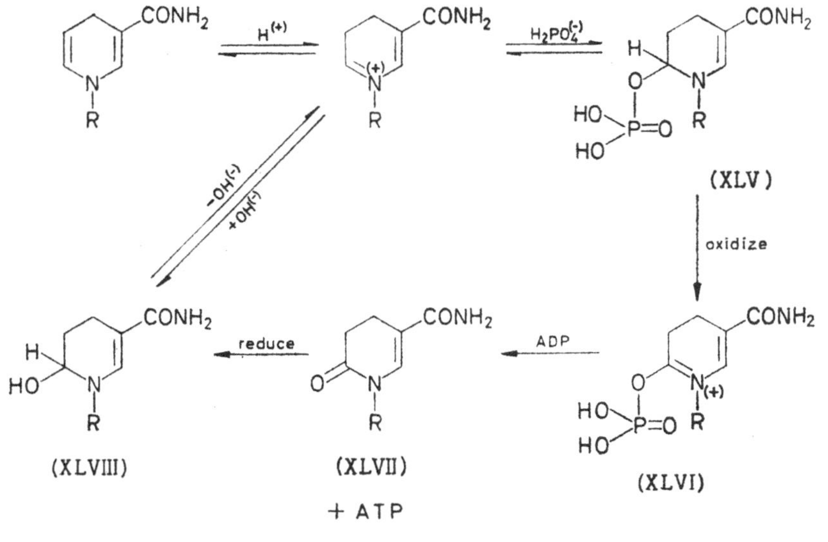

The adduct (XLV), on oxidation, would give (XLVI), an imidoyl phosphate, a class of reagent from which phosphoryl transfer has been amply demonstrated (*vide supra*). This scheme does not operate at the

101

NAD$^+$ level at all, and requires the reduction of an aliphatic amide (XLVII → XLVIII): on both these points it is open to criticism.

Addition of phosphate to NAD$^+$ would lead to an intermediate (L)

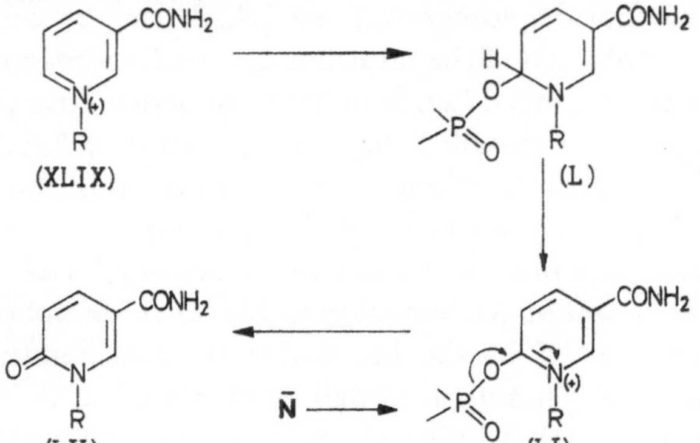

whose oxidation would give (LI): phosphoryl transfer using such a system has already been demonstrated[107]. Which of the two schemes is the more appropriate depends on biochemical observations yet to be made. It should be noted, however, that (XLVI) is merely the protonated form of (L) and that the two schemes differ only in the oxidation levels involved, and not in the feasibility, or otherwise, of phosphoryl transfer. It is likely that reduction of (LII) to regenerate (XLIX)

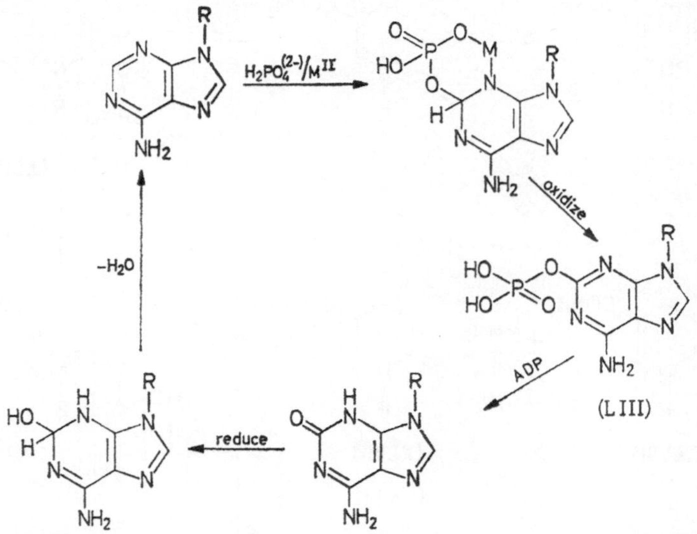

would be easier to accomplish than the corresponding stage (XLVII → XLVIII) in the earlier scheme. Reductions of pyridones directly to pyridines has been observed in a variety of cases[122].

102

An alternative series of reactions in which nicotinamide-adenine dinucleotide participates in virtue of the adenine nucleus has been put forward[109]. The purine ring is susceptible to nucleophilic attack at the 2- and 8-positions, and this process could be assisted to a marked extent by chelation with metals. Oxidation of the adduct so obtained by the next electron carrier in the respiratory chain (a flavoprotein) would give an adenyl phosphate (LIII) closely analogous to the α-pyridyl phosphates[78, 79]. Phosphoryl transfer could then take place leaving the adenyl moiety oxygenated; by reduction and dehydration, the original coenzyme would be regenerated.

In view of the spectroscopic properties of the highly labile, ^{32}P-labelled, NADH derivative mentioned above, and the rapid loss of tritium from its 6-position[121], this last series of reactions seems unlikely. Other proposals to explain oxidative phosphorylation at the NAD/FAD site have been put forward. One[123] involves phosphorylation of the carboxamide group of the nicotinamide residue, whilst another[124] implicates the iso-alloxazine residue of the flavoprotein. These also seem unlikely in view of the experimental evidence cited above.

Phosphoryl Transfer at the Second Site of ATP Synthesis—Both bacteria and mitochondria contain appreciable quantities of quinones (Vitamin K_2—formula below—and the ubiquinones (XL) respectively) whose participation in oxidative phosphorylation has been demonstrated[111, 116, 125].

Substrate phosphorylation can accompany the oxidation of hydroquinone phosphates *in vitro*[28, 86−89] and it has been suggested that such derivatives from Vitamin K_2 and ubiquinone are active participants in phosphoryl transfer in the respiratory chain[115,126−128].

Ultraviolet-irradiation of certain bacterial systems, e.g. *Mycobacterium phlei*, leads to their losing the ability to conduct oxidative phosphorylation[129, 130]. This ability is restored by the addition of Vitamin K_2, or analogues retaining the exocyclic βγ-double bond. Formation of a chromanol (LV) is presumed to be essential since reduction of the βγ-double bond in (LIV) leads to loss of this activity[131].

However, all quinones so far implicated in oxidative phosphorylation *in vivo*, contain a methyl group in the 2-position of the quinonoid ring and a scheme was put forward[115,128] (itself a development of an earlier proposal[132]) which utilizes both these structural features. The scheme provided a mechanism for the activation of inorganic phosphate leading to phosphoryl transfer, and took into account that, during oxidative phosphorylation in mitochondria, no exchange of the quinonoid oxygen

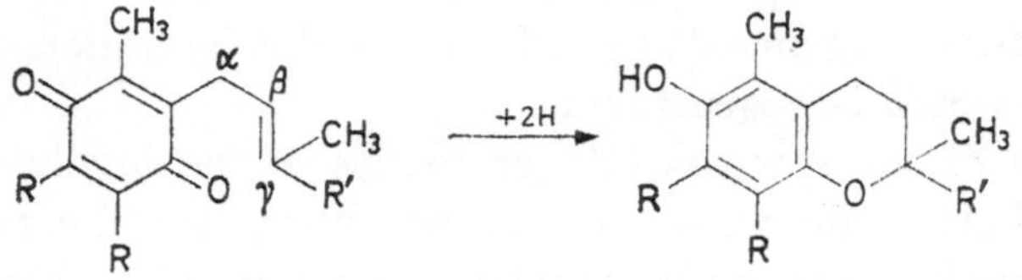

(LIV)   +2H→   (LV)

occurs[133], whereas added inorganic phosphate, labelled with ^{18}O, rapidly loses its label by exchange with water[134].

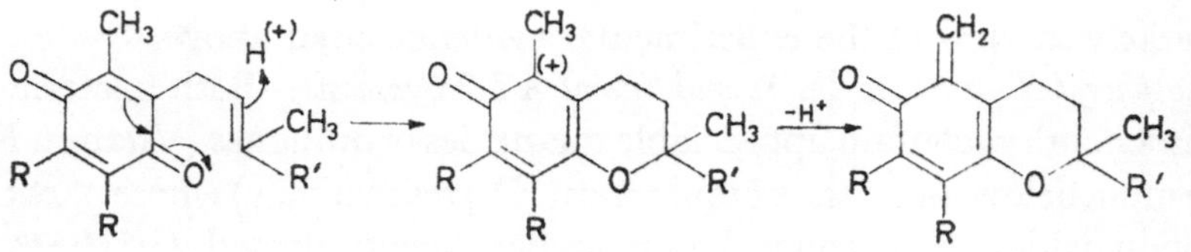

(LIV)   (LVI)

Proton-initiated cyclization of the quinone (LIV) would lead to the quinone-methide (LVI) which, it was proposed, reacted with inorganic

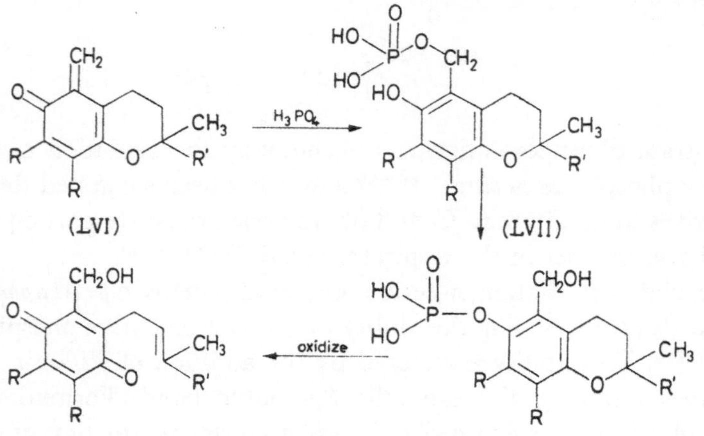

(LVI)   H$_3$PO$_4$→   (LVII)

(LIX)   ←oxidize   (LVIII)

104

phosphate to give the aralkyl phosphate (LVII)[128]. Migration of the phosphoryl residue from the benzylic to the phenolic oxygen atom might then produce the chromanyl phosphate (LVIII), oxidation of which[135] would give (LIX) with concomitant transfer of a phosphoryl group to a substrate, e.g. ADP. Alternatively, hydrolysis of the

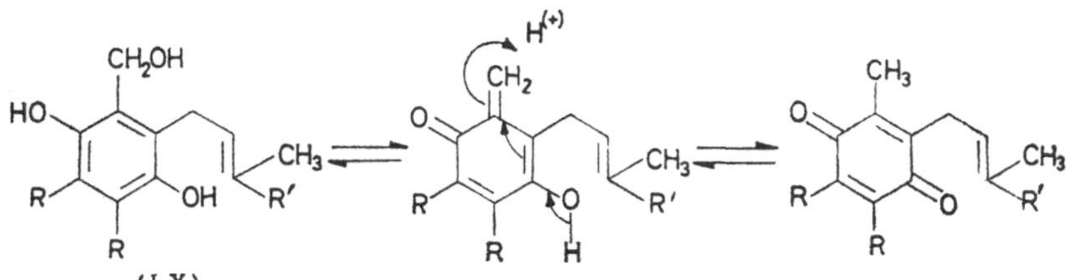

(LX)

phosphoryl group could occur, in which case the inorganic phosphate initially incorporated would have exchanged oxygen with water. In neither case would the quinone/hydroquinone system have exchanged oxygen with the solvent. Reduction of (LIX) to the hydroxymethyl hydroquinone (LX) should be followed by prototropic regeneration of the original 2-methyl quinone, c.f. the formation of duroquinone by hydrolysis of chloromethyl trimethyl hydroquinone[136].

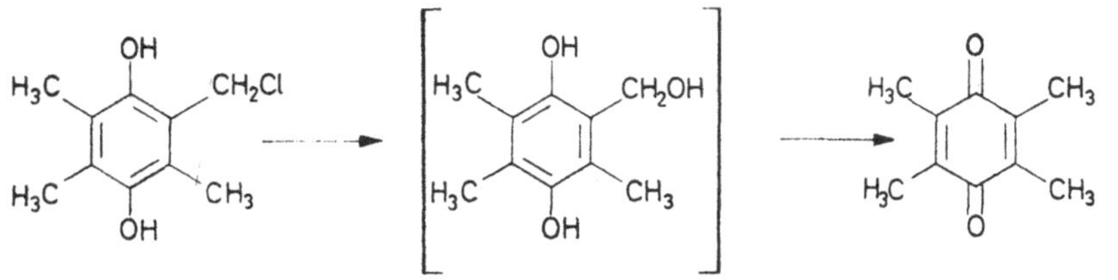

With one important exception, the main features of the scheme are well documented in the chemical literature. The exception lies in the transesterification (LVII) → (LVIII) which requires a phenol to displace an alcohol in a mono-ester of phosphoric acid. This has not been observed in a chemical system.

However, this step is not obligatory since oxidation of (LVII) could lead[137] to (LX). This is a vinylogous acyl phosphate. Ensuing phosphoryl transfer would give the aldehyde (LXI), whose reduction to the hydroxymethyl chromanol (LXII) would allow generation of the quinone-methide (LVI): addition of phosphate would then regenerate (LVII).

105

This last scheme accounts for all the observations yet made using
[18]O-labelled phosphate[133, 134], it separates the phosphoryl transfer
from the oxidative step (which is perhaps desirable on biochemical
grounds[138]), and it implies that the function of the $\beta\gamma$-double bond is
to ensure chroman formation, thereby excluding the prototropic regeneration of the 2-methyl group of the quinone during the cycle.

Studies with Vitamin K[138a, b] and ubiquinones[138c] in which the
2-methyl group was labelled with ^{3}H and ^{14}C indicated no alteration
in the ratio of the two isotopes when the quinones were either incubated
with mitochondria or injected intravenously into rats. These observations have been held to show that quinone methide formation does not
occur during oxidative phosphorylation, as formation of such intermediates would require a diminution in the ratio of ^{3}H to [14]C. Moreover,
in vitro experiments involving duroquinone appeared to support the
absence of hydroged exchange[135, 138e]. However, treatment of duroquinone with secondary aliphatic amines leads to amination[138d] of the
methyl groups.

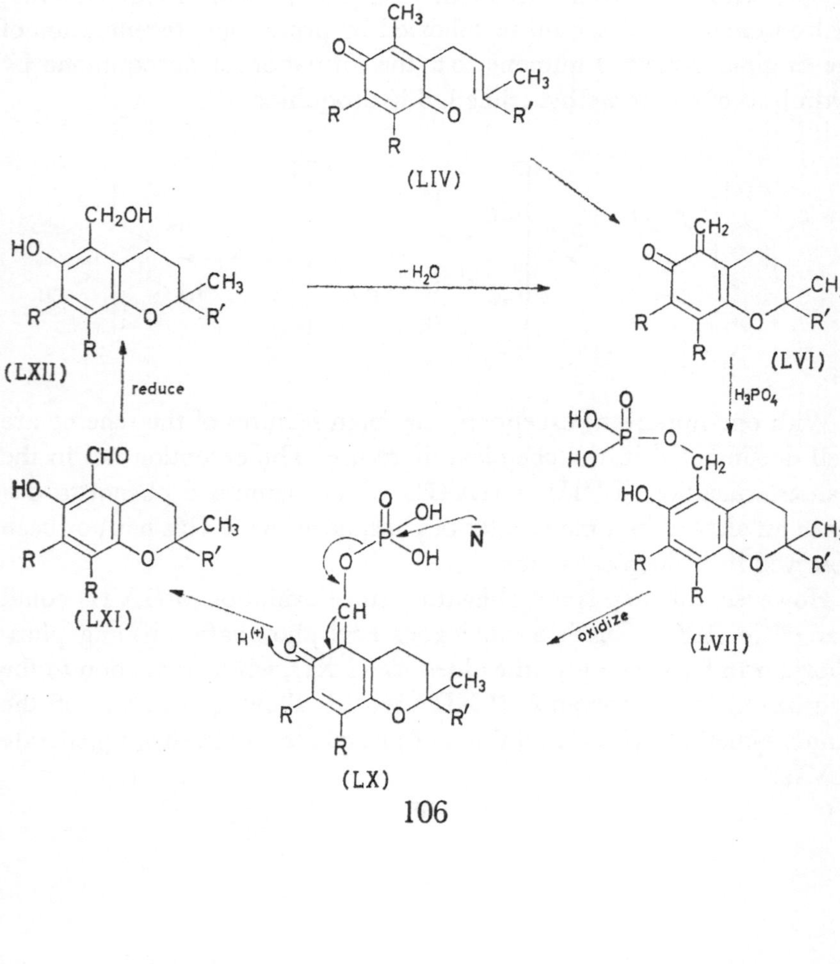

(LIV)

(LXII)

(LVI)

reduce

$-H_2O$

H_3PO_4

(LXI)

(LVII)

oxidize

(LX)

106

The removal of a proton from a methyl quinone gives the ambident anion (LXIII ↔ LXIV) in which the carbon atom of the methyl group could behave either as a nucleophile or an electrophile.

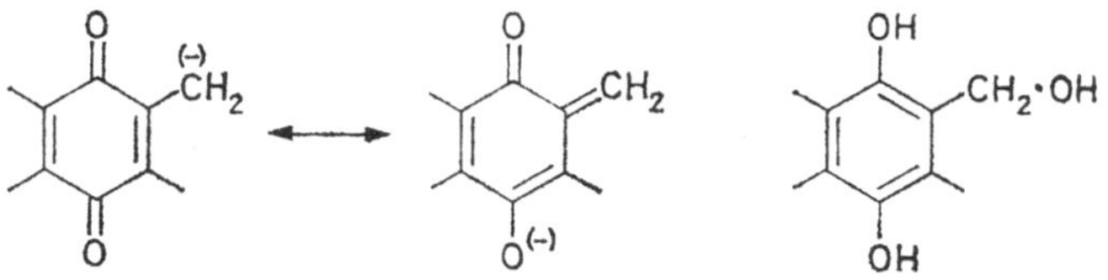

(LXIII)   (LXIV)   (LXV)

In most cases products are related to (LXIV): proton exchange, which has now been demonstrated[138f], requires (LXIII). (LXIV) would react with water to give (LXV) which is analogous to (LXII) required in the cycle presented above. Evidence concerning hydrogen isotope exchange in *recovered* ubiquinones or vitamin K may be irrelevant, since this last-mentioned cycle does not involve the parent quinone.

An additional method of phosphoryl transfer involving the methoxyl group of ubiquinone (XL) is possible. In view of the experimental evidence[94] obtained for haloquinones and tetramethoxy benzoquinone, cited above, nucleophilic attack by phosphate on the quinone ring could lead to phosphoryl transfer. The quinonoid product would then require to be methylated to return to the original ubiquinone. This might involve a redox system, but such a scheme would not provide a specific role for the βγ-double bond, and, in this sense, is less satisfying.

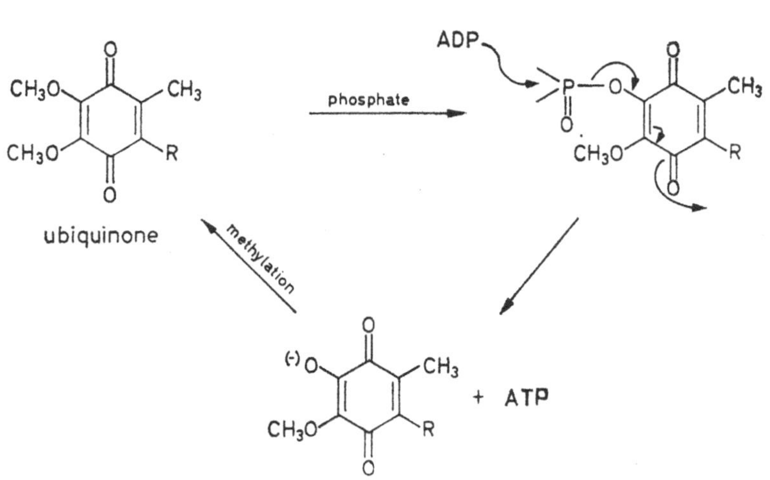

107

Phosphoryl Transfer at the Third Site of ATP Synthesis—The third site of ATP synthesis in the respiratory chain is located between cytochrome *c* and oxygen, probably involving one of the *a* cytochromes. As these are metallo-porphyrins, their mode of action could involve the porphyrin ligand or different valence states of the metal (in this case, iron).

The cytochromes *a*, in contrast to cytochromes *b* and *c*, contain a formyl substituent peripherally attached to the porphyrin nucleus. A scheme for the activation of phosphate involving the formyl group of

where Por = porphyrin

cytochrome a_3 has been proposed[123]. Addition of phosphate to this formyl residue is followed by loss of hydroxyl to give an enol phosphate. This implies transfer of electrons from the nucleus to the side chain and requires that phosphoryl transfer take place from a reduced form of the metalloporphyrin.

A study[139] of the aerial oxidation of ferro-porphyrin in the presence of phosphate and AMP under closely defined conditions showed that ATP could indeed be produced in substantial amounts *only* in those cases in which the porphyrin possessed a formyl substituent. However, whilst the *ferri*-porphyrin system led to ATP formation, the corresponding *ferro*-form was inactive. There was no spectroscopic evidence for the formation of a carboxy-porphyrin from the formyl-substituted systems.

In these sets of experiments the *a* cytochromes are regarded solely as complex aldehydes. It is likely that their true modes of action involve other considerations.

On chelation with a metal ion the porphyrin nucleus loses the two protons attached to nitrogen and the tetradentate ligand must initially be considered as anionic and doubly charged. Also the differences between σ (i.e. coordinate) bonding and π bonding, involving *p* and *d* orbitals, should be considered. Metal complexes fall into two major types[140], viz., *high spin* (more nearly ionic) complexes, e.g. the hydrated ferric ion $Fe(OH_2)_6^{(3+)}$ in which magnetic susceptibility measurements reveal five unpaired electrons, and *low spin* complexes in which several *d* electrons are paired, e.g. $Fe(CN)_6^{(4-)}$.

In $Fe(OH_2)_6^{(3+)}$ the five d electrons of Fe occupy, singly and with their spins parallel, the three t_{2g} orbitals and the two e_g antibonding orbitals whilst the six ligand electron pairs occupy the two bonding e_g, the 4s and the three 4p orbitals. In ferrocyanide, on the other hand, the six electrons of the ferrous ion are paired in three t_{2g} orbitals, and the six electron pairs of the ligands again occupy two e_g orbitals, one 4s and three 4p orbitals[141].

The magnitude of the energy separation between the t_{2g} and e_g orbitals is important since it determines whether or not electrons are unpaired in e_g or paired in t_{2g} orbitals, and this depends on the ligands involved.

With a given metal ion, ligands with lower field strengths (weak-field ligands) tend to form high-spin complexes and strong field ligands lead to low-spin complexes. Ligand field strength increases in the order[142] $Cl^{(-)} < OH^{(-)} < RCO_2^{(-)} < F^{(-)} < H_2O < NH_3 <$ ethylene di-amine $< CN^{(-)}$; no relative value for phosphate is available.

These arguments apply mainly to σ bonding. For the π component, previously non-bonding t_{2g} electron pairs of the metal atom may enter the unoccupied π orbitals of ligands such as $=N-$, CO, O_2, $CN^{(-)}$. The ligand-metal bond is thereby reinforced, the bond length shortened and low spin complexes favoured.

The cytochromes are octahedral porphyrin complexes in which the d_{xz} and d_{yz} orbitals of the metal atom can overlap with vacant π orbitals of the ligands in axial, as opposed to equatorial, positions.

Fe(II) and Fe(III) have, respectively, six and five electrons available from the metal atom. Since there are five d orbitals, a considerable redistribution of electrons (corresponding to the various orbital energies) can result from minor changes in the ligands. Thus, in the octahedral complex of ferrous protoporphyrin with two water molecules in axial positions, four electrons are unpaired[144]: replacement of one water ligand by a molecule of carbon monoxide yields a complex with no unpaired electrons[145].

In porphyrin-protein complexes replacement of a water molecule by a π-bonding ligand easily gives rise to a change from a high to a low-spin complex. Thus, replacement of a single water molecule (a) by oxygen in Fe(II) haemoglobin, or (b) by imidazole in Fe(III) haemi-globin results in a decrease in the number of unpaired electrons, in each case by four units[146].

In cytochromes of the b and c type, it seems that two strong-field ligands in the protein bind the haem iron, one on each side[143].

109

Mammalian cytochrome c is a low-spin complex[147], although on drying, an increase in magnetic moment corresponding to two unpaired electrons has been reported[148]. This implies a change in protein configuration on dehydration. From its spectroscopic properties cytochrome b also appears to be a low spin complex. It is clear that a histidine residue participates in one of the protein–iron bonds in cytochrome c: the second may be associated with a lysine amino—group in the same protein molecule[149].

Cytochrome a_3, like cytochromes b and c, is reduced by an electron passed to it along the electron transport chain, but it subsequently reacts with molecular oxygen (and not another cytochrome) in order to revert to the ferric state. The haem group of cytochrome a_3 must, therefore, have its sixth position available for reaction with O_2.

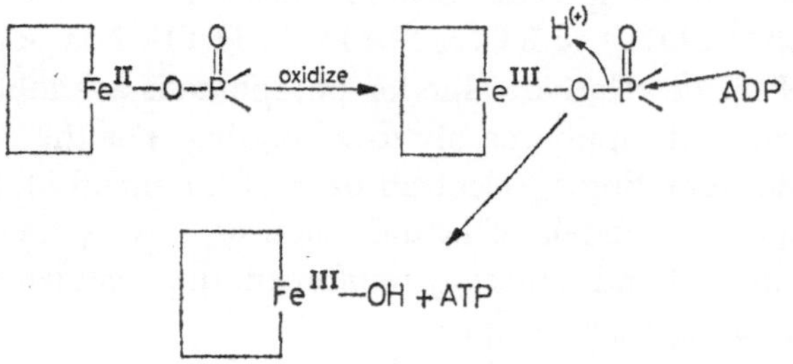

If phosphate is to be activated by an iron-porphyrin system, co-ordination to the metal atom must be followed by a strengthening of the iron–oxygen bond (presumably by oxidation) thereby facilitating

Table II. Redox Potentials of Aqueous Fe(III)/Fe(II) Systems

Ligand/system	pH	E(v)
Fe(OH$_2$)$_6$	1 N acid	0·77
Fe(o-Phenanthroline)$_3$	7	1·1
Cytochrome c	7	0·25
Cytochrome b_6	7	0·05
Fe-Protoporphyrin	7	−0·12
Horseradish peroxidase	7	−0·3

attack at phosphorus with concomitant phosphoryl transfer. The iron atom will then be left with a hydroxyl group (or water molecule) in the octahedral position to which the phosphate group had been originally

110

co-ordinated. Since the oxygen atom of hydroxyl is more nucleophilic*
than an oxygen atom in any of the anionic forms of phosphate, oxidation to the ferric state should favour phosphorus–oxygen cleavage. This
implies that the Fe(III) state is the active one, and experimental
evidence to this effect has already been obtained[139].

The redox potential of ferrous-ferric systems varies remarkably with
changes in the octahedral framework (*Table II*)[150] and it is clear that
phosphoryl transfer could proceed according to the above mechanism
given appropriate ligands.

POSTSCRIPT

'*Our* Phosphorus *is a subject that occupies much the Thoughts and Fancies
of some Alchymists, who work on Microcosmical substances: and out of it they
promise themselves Golden Mountains*'[151].

REFERENCES

[1] KHORANA, H. G. *Some Recent Developments in the Chemistry of Phosphate
Esters of Biological Interest.* Wiley, New York, 1961
[2] BROWN, D. M. *Advances in Organic Chemistry: Methods and Results* (Raphael,
Taylor and Wynberg, Eds). Vol. 3, p. 76, Interscience, New York, 1963
[3] KIRBY, A. J. and WARREN, S. G. *The Organic Chemistry of Phosphorus*, Elsevier,
Amsterdam, 1967
[4] HUTCHINSON, D. W. *Nucleotides and Coenzymes*, Methuen, London, 1965
[5] CRUICKSHANK, D. W. J. *J. chem. Soc.* (1961) 5486
[6] McOMIE, J. F. W. *Advances in Organic Chemistry: Methods and Results*
(Raphael, Taylor and Wynberg, Eds). Vol. 3, p. 191, Interscience, New
York, 1963
[7] VERNON, C. A. *Chem. Soc. Publ. No. 8* (1957) 17
[8] CRAIG, D. P., MACCOLL, A., NYHOLM, R. S., ORGEL, L. E. and SUTTON,
L. E. *J. chem. Soc.* (1954) 332
[9] CRAIG, D. P. and MAGNUSSON, E. A. *J. chem. Soc.* (1956) 4895
[10] TRIPPETT, S. *Quart. Rev. chem. Soc.* 17 (1963) 406
[11] BESTMANN, H. J. quoted in Hudson, R. F. *Pure appl. Chem.* 9 (1964) 371
[12] DOERING, W. VON E. and HOFFMANN, A. K. *J. Amer. chem. Soc.* 77 (1955)
521
[13] HOBBS, E., CORBRIDGE, D. E. C. and RAISTRICK, B. *Acta Cryst.* 6 (1953) 621
[14] MASON, H. S. and TODD, A. R. *J. chem. Soc.* (1951) 2267
[15] WEIMANN, G. and KHORANA, H. G. *J. Amer. chem. Soc.* 84 (1962) 4329
[16] CLARK, V. M., KIRBY, G. W. and TODD, SIR ALEXANDER. *J. chem. Soc.*
(1957) 1497
[17] MOFFATT, J. G. and KHORANA, H. G. *J. Amer. chem. Soc.* 83 (1961) 649
[18] KHORANA, H. G. and MOFFATT, J. G. *J. Amer. chem. Soc.* 83 (1961) 663
[19] STOKES, H. N. *Amer. chem. J.* 15 (1893) 198

* At least by reference to a proton in water; the pKa of H_2O is greater than that of the
third dissociation of H_3PO_4. Changes in dielectric constant might affect this.

20 SKROWACZEWSKA, Z. and MASTERLERZ, P. *Roczniki Chem.* 29 (1955) 415
21 HAMER, N. K. *J. chem. Soc.* (1965) 46
21a HAMER, N. K. *J. chem. Soc.* (C) (1966) 404
22 HAMER, N. K. *J. chem. Soc.* (1965) 2731
23 PAULING, L. *The Nature of the Chemical Bond*, 3rd edn, p. 93, Cornell University Press, Ithaca, N.Y., 1960
24 KIRBY, G. W., *Ph.D. Thesis, Cambridge University*, 1958
25 CLARK, V. M., LORD TODD and WARREN, S. G. *Biochem. Z.* 338 (1963) 591
26 CLARK, V. M., HUTCHINSON, D. W., KIRBY, A. J. and WARREN, S. G. *Angew. Chem.* 3 (1964) 678
27 LARSSON, L. *Svensk kem. Tidskr.* 70 (1958) 405
28 CLARK, V. M., HUTCHINSON, D. W., KIRBY, G. W. and TODD, SIR ALEXANDER. *J. chem. Soc.* (1961) 715
29 MAYNARD, J. A. and SWAN, J. M. *Proc. chem. Soc.* (1963) 61
30 MAYNARD, J. A. and SWAN, J. M. *Aust. J. Chem.* 16 (1963) 596
31 DI SABATO, G. and JENCKS, W. P. *J. Amer. chem. Soc.* 83 (1961) 4400
32 CONANT, J. B. and COOK, A. A. *J. Amer. chem. Soc.* 42 (1920) 830
33 CONANT, J. B. and POLLACK, S. M. *J. Amer. chem. Soc.* 43 (1921) 1665
34 CONANT, J. B. and COYNE, B. B. *J. Amer. chem. Soc.* 44 (1922) 2530
35 MEISTERS, A. and SWAN, J. M. *Aust. J. Chem.* 18 (1965) 155
36 KREUTZKAMP, N. and KAYSER, H. *Chem. Ber.* 89 (1956) 1614
37 MEISTERS, A. and SWAN, J. M. *Aust. J. Chem.* 16 (1963) 725
38 ZIOUDROU, C. *Tetrahedron*, 18 (1962) 197
39 HALMANN, M., LAPIDOT, A. and SAMUEL, D. *J. chem. Soc.* (1960) 4672
40 HALMANN, M. and LAPIDOT, A. *J. chem. Soc.* (1960) 419
41 BUNYAN, P. J. and CADOGAN, J. I. G. *J. chem. Soc.* (1962) 1304
42 HAMER, N. K. *J. chem. Soc.* (1964) 1961
43 DI SABATO, G. and JENCKS, W. P. *J. Amer. chem. Soc.* 83 (1961) 4393
44 BENTLEY, R. *J. Amer. chem. Soc.* 71 (1949) 2765
45 HALMANN, M., LAPIDOT, A. and SAMUEL, D. *J. chem. Soc.* (1962) 1944
46 CRAMER, F. and WINTER, M. *Chem. Ber.* 92 (1959) 2761
47 KHORANA, H. G. and VIZSOLYI, J. P. *J. Amer. chem. Soc.* 81 (1959) 4660
48 WOODWARD, R. B., OLOFSON, R. A. and MAYER, H. *J. Amer. chem. Soc.* 83 (1961) 1010
49 CRAMER, F., NEUNHOFFER, H., SCHEIT, K. H., SCHNEIDER, G. and TENNIGKEIT, J. *Angew. Chem.* 1 (1962) 387
50 JACOB, T. M. and KHORANA, H. G. *J. Amer. chem. Soc.* 86 (1964) 1630
51 STOCKX, J. *Bull. Soc. chim. belges.* 70 (1961) 595
52 WITTMANN, R. *Chem. Ber.* 96 (1963) 2116
53 KIRBY, A. J. and VARVOGLIS, A. G. *J. Amer. chem. Soc.* 89 (1967) 415
54 BUNTON, C. A., FENDLER, E. J. and FENDLER, J. H. *J. Amer. chem. Soc.* 89 (1967) 1221
55 CLARK, V. M. and WARREN, S. G. *Nature*, Lond. 199 (1963) 657
56 BADDILEY, J., BUCHANAN, J. G. and LETTERS, R. *J. chem. Soc.* (1956) 2812
57 RATHLEV, T. and ROSENBERG, T. *Archs Biochem. Biophys.* 65 (1956) 319
58 HELLMANN, H., LINGENS, F. and BURKHARDT, H. *J. Chem. Ber.* 91 (1958) 2290
59 GOLDMAN, L., MARSICO, J. W. and ANDERSON, G. W. *J. Amer. chem. Soc.* 82 (1960) 2969
60 CRAMER, F., SCHALLER, H. and STAAB, H. A. *Chem. Ber.* 94 (1961) 1612
61 LICHTENTHALER, F. W. *Chem. Rev.* 61 (1961) 607
62 CRAMER, F. and GÄRTNER, K. G. *Chem. Ber.* 91 (1958) 704

[63] CRAMER, F. and WITTMANN, R. *Angew. Chem.* 72 (1960) 628

[64] KREUTZKAMP, N. and KAYSER, H. *Liebigs Ann.* 609 (1957) 39

[65] ARENS, J. F. and DOORNBOS, T. *Rec. Trav. chim. Pays-Bas* 74 (1955) 79

[66] WASSERMAN, H. H. and COHEN, D. *J. org. Chem.* 29 (1964) 1817

[67] KHORANA, H. G. and TODD, A. R. *J. chem. Soc.* (1953) 2257

[68] VILSMEIER, A. and HAACK, A. *Ber. dtsch. chem. Ges.* 60 (1927) 119

[69] BREDERECK, H., GOMPPER, R., KLEMM, K. and REMPFER, H. *Chem. Ber.* 92 (1959) 837

[70] CRAMER, F. and WEIMANN, G. *Chem. Ber.* 94 (1961) 996

[71] CREMLYN, R. J. W., KENNER, G. W. and TODD, SIR ALEXANDER. *J. chem. Soc.* (1960) 4511

[72] FOX, R. B. and BAILEY, W. J. *J. org. Chem.* 26 (1961) 2542

[73] KENNER, G. W., REESE, C. B. and TODD, SIR ALEXANDER. *J. chem. Soc.* (1958) 546

[74] KENNER, G. W., TODD, SIR ALEXANDER and WEBB, R. F. *J. chem. Soc.* (1956) 1231

[75] ZAORAL, M. and ARNOLD, Z. *Tetrahedron Ltrs.* (1960) 9

[76] CRAMER, F., RIITNER, S., REINHARD, W. and DESAI, P. *Chem. Ber.* 99 (1966) 2252

[77] CRAMER, F. and WINTER, M. *Chem. Ber.* 94 (1961) 989

[78] KAMPE, W. *Chem. Ber.* 98 (1965) 1031

[79] KAMPE, W. *Chem. Ber.* 98 (1965) 1038

[80] SCHEIT, K. H. and KAMPE, W. *Chem. Ber.* 98 (1965) 1045

[81] CRAMER, F. and WITTMANN, R. *Angew. Chem.* 73 (1961) 220

[82] MUKAIYAMA, T., HATA, T. and MITSONOBU, O. *J. org. Chem.* 27 (1962) 1815

[83] BUNTON, C. A., LLEWELLYN, D. R., OLDHAM, K. G. and VERNON, C. A. *J. chem. Soc.* (1958) 3588

[84] WOLMAN, Y., GALLOP, P. M. and PATCHORNIK, A. *J. Amer. chem. Soc.* 83 (1961) 1263

[85] BROWN, D. M., FLINT, J. A. and HAMER, N. K. *J. chem. Soc.* (1964) 326

[86] WIELAND, T. and PATTERMANN, F. *Chem. Ber.* 92 (1959) 2917

[87] CLARK, V. M., HUTCHINSON, D. W. and TODD, SIR ALEXANDER. *J. chem. Soc.* (1961) 722

[88] LAPIDOT, A. and SAMUEL, D. *Biochim. biophys. Acta.* 65 (1962) 164

[89] LAPIDOT, A. and SAMUEL, D. *J. Amer. chem. Soc.* 86 (1964) 1886

[90] EGGINS, B. R. and HUTCHINSON, D. W. Unpublished results

[91] DÜRCKHEIMER, W. and COHEN, L. A. *Biochemistry* 3 (1964) 1948

[92] ANDREWS, K. J. M. *J. chem. Soc.* (1961) 1808

[93] COREY, E. J. and KÖNIG, H. J. *J. Amer. chem. Soc.* 84 (1962) 4904

[94] CLARK, V. M., HUTCHINSON, D. W. and ROSCHNIK, R. K. *Chem. & Ind.* (1965) 135

[95] ROSCHNIK, R. K., *Ph.D. Thesis*, Cambridge University, 1965

[96] LYONS, A. R., *M.Sc. Thesis*, Univ. of Warwick, 1966

[97] WITTMANN, R. *Chem. Ber.* 96 (1963) 771

[98] BUCKLEY, D., DUNSTAN, S. and HENBEST, H. B. *J. chem. Soc.* (1957) 4880

[99] EISTERT, B. and BOCK, G. *Chem. Ber.* 92 (1959) 1239

[100] LINN, B. O., TRENNER, N. R., ARISON, B., WESTON, R. G., SHUNK, C. H. and FOLKERS, K. *J. Amer. chem. Soc.* 82 (1960) 1647

[101] WALLENFELS, K. and BACHMANN, G. *Angew. Chem.* 73 (1961) 142

[102] MOORE, H. W. and FOLKERS, K. *J. Amer. chem. Soc.* 87 (1965) 1409

[103] MARXER, A. *Helv. Chim. Acta.* 40 (1957) 502

104 Morton, R. A. *Quinones in Electron Transport* (Wolstenholme and O'Connor, Eds), p. 5, Churchill, London, 1961
105 Crane, F. L. *Quinones in Electron Transport* (Wolstenholme and O'Connor, Eds), p. 36, Churchill, London, 1961
106 Redfearn, E. R. *Vitamins and Hormones* 24 (1966) 456
107 Clark, V. M., Hutchinson, D. W. and Wilson, D. E. *Angew. Chem.* 77 (1965) 259
108 Griffiths, D. E. *Federation Proc.* 22 (1963) 1064
109 Barltrop, J. A., Grubb, P. W. and Hesp, B. *Nature, Lond.*, 199 (1963) 759
110 Lightbown, J. W. and Jackson, F. L. *Biochem. J.* 63 (1956) 130
111 Asano, A. and Brodie, A. F. *J. biol. Chem.* 239 (1964) 4280
112 Fruton, J. S. and Simmonds, S. *General Biochemistry*, 2nd edn, p. 521, Wiley, New York, 1959
113 Lehninger, A. L. and Wadkins, C. L. *Ann. Rev. Biochem.* 31 (1962) 47
114 Griffiths, D. E. *Essays in Biochemistry* (Campbell and Greville, Eds). Vol. 1, p. 91, Academic Press, New York, 1965
115 Lederer, E. and Vilkas, M. *Vitamins and hormones* 24 (1966) 409
116 Lester, R. L. and Fleischer, S. *Arch. biochem. Biophys.* 80 (1959) 470
117 Chance, B. and Redfearn, E. R. *Biochem. J.* 80 (1961) 632
118 Green, D. E. *Quinones in Electron Transport* (Wolstenholme and O'Connor, Eds), p. 130, Churchill, London, 1961
119 Purvis, J. L. *Biochim. biophys. Acta.* 38 (1960) 435
120 Pinchot, G. B. *Proc. Nat. Acad. Sci. U.S.* 46 (1960) 929
121 Griffiths, D. E. *Abstr. VIth Internat. Congr. Biochem.*, New York, 1964, p. 766
122 De Mayo, P. and Rigby, W. *Nature*, Lond. 166 (1950) 1075
123 Glahn, P. and Nielsen, S. O. *Nature*, Lond. 183 (1959) 1578
124 Lindberg, O., Grabe, B. O., Löw, H., Siekevitz, P. and Ernster, L. *Acta chem. Scand.* 12 (1958) 598
125 Colpa-Boonstra, J. B. and Slater, E. C. *Biochim. biophys. Acta.* 27 (1958) 122
126 Clark, V. M. and Todd, Sir Alexander. *Quinones in Electron Transport* (Wolstenholme and O'Connor, Eds), p. 190, Churchill, London, 1961
127 Clark, V. M. *Mechanismen Enzymatischer Reaktionen*, p. 276, Springer-Verlag, Berlin, 1964
128 Vilkas, M. and Lederer, E. *Experientia.* 18 (1962) 546
129 Brodie, A. F. and Ballantine, J. *J. biol. Chem.* 235 (1960) 226
130 Brodie, A. F. and Ballantine, J. *J. biol. Chem.* 235 (1960) 232
131 Russell, P. J. and Brodie, A. F. *Quinones in Electron Transport* (Wolstenholme and O'Connor, Eds), p. 205, Churchill, London, 1961
132 Chmielewska, I. *Biochim. biophys. Acta.* 39 (1960) 170
133 Snyder, C. D. and Rapoport, H. *J. Amer. chem. Soc.* 89 (1967) 1269
134 Cohn, M. *J. biol. Chem.* 201 (1953) 735
135 Scott, P. M. *J. biol. Chem.* 240 (1965) 1374
136 Smith, L. I. and Carlin, R. B. *J. Amer. chem. Soc.* 64 (1942) 524
137 Wagner, A. F., Lusi, A., Shunk, C. H., Linn, B. O., Wolf, D. E., Hofmann, C. H., Erickson, R. E., Arlson, B., Trenner, N. R. and Folkers, K. *J. Amer. chem. Soc.* 85 (1963) 1534
138 Racker, E. *adv. Enzymol.* 23 (1961) 323
138a Horth, C. E., McHale, D., Jeffries, L. R., Price, S. A., Diplock, A. T. and Green, J. *Biochem. J.* 100 (1966) 424
138b Snyder, C. D., Di Mari, S. J. and Rapoport, H. *J. Amer. chem. Soc.* 88 (1966) 3868

138c PARSON, W. W. and RUDNEY, H. *Biochemistry* 5 (1966) 1013

138d CAMERON, D. W., SCOTT, P. M. and LORD TODD. *J. chem. Soc.* (1964) 42

138e LAPIDOT, A., SILVER, B. L. and SAMUEL, D. *J. biol. Chem.* 241 (1966) 5537

138f CLARK, V. M., HUTCHINSON, D. W. and WILSON, R. G. *Nature*, Lond. in press

139 BRINIGAR, W. S. and WANG, J. H. *Proc. natn. Acad. Sci. U.S.* 52 (1964) 699

140 COTTON, F. A. and WILKINSON, G. *Advanced Inorganic Chemistry*, 2nd edn, p. 671, Interscience, New York, 1966

141 ORGEL, L. E. *An Introduction to Transition Metal Chemistry*, 3rd edn, Ch. 3, Methuen, London, 1966

142 BASOLO, F. and PEARSON, R. G. *A Study of Metal Complexes in Solution*, p. 121, Wiley, New York, 1958

143 FALK, J. E. *Porphyrins and Metalloporphyrins*, p. 54, Elsevier, Amsterdam, 1964

144 PAULING, L. and CORYELL, C. D. *Proc. Nat. Acad. Sci. U.S.* 22 (1936) 159

145 WANG, J. H., NAKAHARA, A. and FLEISCHER, E. B. *J. Amer. chem. Soc.* 80 (1958) 1109

146 CORYELL, C. D., STITT, F. and PAULING, L. *J. Amer. chem. Soc.* 59 (1937) 633

147 THEORELL, H. *J. Amer. chem. Soc.* 63 (1941) 1820

148 LUMRY, R., SOLBAKKEN, A., SULLIVAN, J. and REVERSON, L. H. *J. Amer. chem. Soc.* 84 (1962) 142

149 MARGOLIASH, E., FROHWIRT, N. and WIENER, E. *Biochem. J.* 71 (1959) 559

150 CALVIN, M. *Rev. Pure appl. Chem.* 15 (1965) 1

151 HANCKEWITZ, A. G. *Phil. Trans. R. Soc.* 38 (1733) 67

115

5

ACIDIC HYDROCARBONS

H. Fischer and D. Rewicki

INTRODUCTION AND SCOPE	116
DEFINITION OF ACID STRENGTH	
Kinetic and equilibrium acidity	117
Intrinsic acidity and thermodynamic acidity scales	118
MOLECULAR STRUCTURE AND CH-ACIDITY	
Molecular energies and intrinsic acidity	122
HMO energies and intrinsic acidity	123
HMO energies and experimental acid strength	126
MEASUREMENT OF CH-ACIDITY	
Measurement of K_r-values by equilibrium methods	128
Measurement of K_r-values by kinetic methods	134
pK-values of acidic hydrocarbons referred to water	135
INTERRELATION OF THEORETICAL AND EXPERIMENTAL RESULTS	
Comparison of uv-spectra of anions of acidic hydrocarbons with calculations	142
Comparison of p_wK-values of acidic hydrocarbons with calculations	144
SYNTHESES AND REACTIONS OF ACIDIC HYDROCARBONS AND CYANOCARBON ACIDS	
Syntheses	148
Reactions	151

INTRODUCTION AND SCOPE

WHILE the substitution of hydrogen with metals in certain hydrocarbons (acetylene[1], phenylacetylene[2], triphenylmethane[3], diphenylmethane[4], cyclopentadiene[5], fluorene[6]) has been a known reaction for 100 years, it was not until the turn of the century that 'acidity' was made responsible for this phenomenon. Unsuccessful attempts to determine an acid dissociation constant of acetylene in water date from the same time[7]. Later, chemists became familiar with the concept of CH-acidity mainly in connection with investigations of tautomeric equilibria. Around 1930, differences in acid strength of some acidic hydrocarbons were noticed[8-10]. A few years later CONANT and WHELAND[11] and McEWEN[12] set up a semiquantitative pK-scale for a number of hydrocarbons. It is only in recent years that noticeable progress has been made concerning synthesis and more exact pK measurement of acidic hydrocarbons.

The emphasis of this chapter lies on hydrocarbon acids of pronounced acidity. CH-acidity caused by halogeno-, nitro-, oxo- or sulpho-substituents is not considered. However, the so-called cyanocarbon acids, some of which surpass mineral acids in acid strength, are included.

The acidity of hydrocarbons is of importance for their reactive behaviour, but still more importance attaches to the theoretical interpretation of the acidity. There has been a theoretical interest in such hydrocarbons since the early days of quantum chemistry[13, 14]. This is due to the fact that acidic hydrocarbons are nearly always π-electron molecules and π-electron theory has been continuously developed since the advent of Hückel theory in 1931.

One of the main purposes of this chapter is to point out the interrelationship between theory and experiment concerning the acidity of hydrocarbons. To this end one must work out what is measured on the one hand and what is calculated, using suitable models, on the other. This is attempted in the following two sections.

DEFINITION OF ACID STRENGTH

Kinetic and Equilibrium Acidity

The dissociation of a proton acid may be treated in two different ways:

(i) by considering the rate of a reaction involving proton transfer (kinetic or dynamic acidity), measured by a rate constant k;

(ii) by considering the position of the equilibrium between the acid and an appropriate base (equilibrium, thermodynamic or static acidity), measured by an equilibrium constant K.

A relationship which interrelates kinetic and thermodynamic acidity was proposed by BRÖNSTED and PEDERSEN[15] based on work on the acid- or base-catalysed decomposition of nitramide. If k denotes the rate constant of an acid catalysed reaction and K the thermodynamic dissociation constant of the acid used (both measured in the same acid-base system) one has the relation

$$\log k = \alpha \log K + \log G \qquad (0 < \alpha < 1) \qquad (1)$$

G and α are expected to be constant for a particular reaction using different acids as catalysts. It turns out[16] that eqn (1) does not hold generally unless a series of very similar acids is used. Equation (1) is entirely invalid if for example carbon- and oxygen-acids are compared with one another[17].

The basis of the Brönsted relationship has been critically examined by EIGEN[18] in connection with an extensive study of the rates of protolytic reactions. Since many catalysed reactions are determined kinetically by a single protolytic reaction between the catalyst and the substrate, eqn (1) should also hold for the rate constants of simple acid-base reactions. For several of such proton transfer reactions Eigen[18] has studied the dependence of log K on the pK-difference between proton donor and acceptor over a wide pK-range. As expected, theoretically no linear relationship was apparent. Furthermore, the relationships actually obtained depend on the type of acid investigated. According to Eigen, the success of linear approximations corresponding to eqn (1) was due to the fact that by classical kinetic methods one can only measure rate constants which are substantially smaller than the limiting values. In this case linear approximations may extend over a relatively wide pK-range.

Obviously kinetic and thermodynamic acidity are in general quite different things. Thus the differentiation between OH-, NH- and CH-acidity is only meaningful if kinetic acidity is meant[19]. From a thermodynamic point of view there is, in principle, no difference between OH-, NH- and CH-acids.

Intrinsic Acidity and Thermodynamic Acidity Scales

BRÖNSTED[20] has defined an acid RH as a compound showing a certain tendency to give up a proton. A measure of the *intrinsic acidity* of RH would be the equilibrium constant K_{abs} (absolute acidity constant) of the reaction

$$RH \rightleftharpoons R^- + H^+ \tag{2}$$

defined as

$$K_{abs} = \frac{a'_{R^-} \cdot a'_{H^+}}{a'_{RH}} \tag{3}$$

where a'_X denotes the absolute activity of the species X at equilibrium. The absolute activities may be defined such that they become identical with mole fractions in the ideal gas phase where no interactions between the particles occur. This definition renders K_{abs} independent of any medium.

Protolytic reactions involving equilibria like eqn (2) are usually investigated in solution, where interactions between all the dissolved

118

particles and interactions between the solvent and the solute occur. The absolute activities would allow for all these interactions, but absolute activities cannot be measured. In solution we can only determine concentrations or conventional activities of *solvated particles*. This suggests measurement of the acid strength of RH in a particular solvent by means of the thermodynamic dissociation constant $_sK$ defined in analogy to K_{abs} by

$$_sK = \frac{a_{R^-} \cdot a_{H^+}}{a_{RH}} = \frac{[R^-] \cdot [H^+]}{[RH]} \cdot \frac{f_{R^-} \cdot f_{H^+}}{f_{RH}} \tag{4}$$

where a_X and $[X]$ denote conventional activities and concentrations of solvated particles, respectively, and f_X is the conventional activity coefficient. In this formulation 'solvation' comprises a possible protolytic reaction of the proton with the solvent. Thus $_sK$-values will strongly depend on the solvation power of the solvent and in particular on its tendency to accept protons.

For weak acids like hydrocarbons the $_sK$-values will often be immeasurably small. In this case it is convenient to refer the acid strength of RH to a second acid (BH^+ say) of comparable strength. This is done by measuring the equilibrium constant K_r of the reaction

$$RH + B \rightleftharpoons R^- + BH^+ \tag{5}$$

defined by

$$K_r = \frac{_sK^{RH}}{_sK^{BH^+}} = \frac{a_{R^-} \cdot a_{BH^+}}{a_{RH} \cdot a_B} = \frac{[R^-] \cdot [BH^+]}{[RH] \cdot [B]} \cdot \frac{f_{R^-} \cdot f_{BH^+}}{f_{RH} \cdot f_B} \tag{6}$$

Obviously the K_r-values of a series of acids referred to a standard acid are proportional to the corresponding $_sK$-values in the solvent used. $_sK$ is related to K_{abs} by means of absolute activity coefficients f':

$$K_{abs} = _sK \cdot \frac{f'_{R^-} \cdot f'_{H^+}}{f'_{RH}} \tag{7}$$

According to the above definition of $_sK$ the absolute activity coefficients f' contain all the medium effects, which will be a function of the individual character of both the dissolved acid and the solvent. Thus $_sK$ (or K_r) does not necessarily reflect the intrinsic acidity of an acid (or the relative intrinsic acidities of two acids) as expressed by K_{abs}. The $_sK$-values may vary from solvent to solvent in an unpredictable manner

119

such that even an inversion of the relative acid strength of two acids may result when the solvent is changed [21]. In view of the construction of an acidity scale, independent of the solvent used, $_sK$ turns out to be a rather arbitrary measure of acid strength.

To overcome this difficulty one must attempt to compare acid strengths under conditions where the ratio f'_{R^-}/f'_{RH} in eqn (7) is approximately constant. Two possibilities may be envisaged: (i) only structurally very similar acids are compared; (ii) the interaction of the solvent with the solute is minimized.

(i) A reasonable assumption is that a series of structurally very similar acids and the series of its conjugated bases show a similar inter- action with the solvent. If this is true, the $_sK$-values of these acids should parallel the corresponding series of K_{abs}-values or the K_r-values of such acids should be the same in all solvents. An important example where the condition of similar interaction with the solvent should be fulfilled is provided by the tautomeric hydrocarbon acids (p. 147).

An important point of this alternative is that there are no restrictions as to the nature of the solvents to be used: since the conventional activities in eqns (4) and (6) are, in general, not measurable by direct methods, one is forced to rely on concentrations. This can be justified if the interactions between the dissolved particles themselves can be suppressed (the conventional activity coefficients in eqns (4) and (6) would then tend to unity). Since ionic species are always involved in protolytic reactions, this can be achieved only by working in high dilution in solvents of high dielectric constants.

(ii) It may be assumed that in non-polar solvents like benzene or petroleum ether where the interactions between the dissolved particles and the solvent are in general weak and unspecific, the absolute activity coefficients in eqn (7) will tend to unity. This should allow the com- parison of acid strength, even of structurally dissimilar acids, and the K_r-values thus obtained should agree with the corresponding relative K_{abs}-values.

In this case, due to the low dielectric constant of the solvent, strong interactions between the ionic species will occur such that the protolytic reaction no longer corresponds to eqn (5). The actual situation has to be described by an equilibrium involving ion pairs[22]:

$$RH + B \overset{K_i}{\rightleftharpoons} R^-BH^+ \overset{K_d}{\rightleftharpoons} R^- + BH^+ \qquad (8)$$

Now the overall equilibrium constant K_r of the protolytic reaction (8) is composed of the ionization constant K_i and of the dissociation constant K_d of the ion pairs:

$$K_r = \frac{{}_sK^{RH}}{{}_sK^{BH^+}} = \frac{a_{R^-} \cdot a_{BH^+}}{a_{RH} \cdot a_B} = K_i \cdot K_d \qquad (9)$$

Suppose the ion pairs are almost undissociated and the activities may be replaced by concentrations to a first approximation, then the ionization constant K_i is the only measurable quantity (by an optical method). Comparing two acids R_1H and R_2H using BH^+ as standard acid, the ratio of K_i-values given by

$$\frac{K_i^{R_1H}}{K_i^{R_2H}} = \frac{[R_1^-BH^+] \cdot [R_2H]}{[R_2^-BH^+] \cdot [R_1H]}$$

$$= \frac{[R_1^-] \cdot [R_2H] K \cdot {}_d^{R_2H}}{[R_2^-] \cdot [R_1H] \cdot K_d^{R_1H}} \qquad (10)$$

$$= \frac{{}_sK^{R_1H} \cdot K_d^{R_2H}}{{}_sK^{R_2H} \cdot K_d^{R_1H}}$$

is not closely related to the ratio of the thermodynamic dissociation constants ${}_sK$ unless the dissociation constants K_d of the two ion pairs are equal. But there is no reason to expect this equality to hold generally. Furthermore, the ratio of the K_d-values may depend on the nature of the base B[23]. Again, it is reasonable to assume such complicating features are absent in the case of structurally similar acids. The ratio of the ionization constants K_i will then parallel the corresponding ratios of ${}_sK$- and K_{abs}-values.

The foregoing discussion demonstrates that a pK-scale will reflect the intrinsic acidity if all the acids can be assumed to behave similarly in any solvent. In this sense one might consider all hydrocarbon acids to belong to the same type and NH- and OH-acids to different types.

MOLECULAR STRUCTURE AND CH-ACIDITY

In this section it is desired to establish a connection between theoretical energy calculations on a molecular level and acidity as expressed by the constants K_{abs} or ${}_sK$. Acidic hydrocarbons discussed in this article are always π-electron molecules for which the most apparent change associated with carbanion formation occurs with the π-electrons. This suggests that the change in π-electron energy may play an essential role in determining the acid strength of such hydrocarbons, an exception being compounds the acidity of which is due to an increased *s*-character of the CH-bond in question. Thus a strong increase in acidity is observed in the series ethane, cyclopropane, ethylene and acetylene[24]. The acid strength of acetylenes is apparently not appreciably affected by either inductive or mesomeric effects[25].

Molecular Energies and Intrinsic Acidity

The equilibrium constant K_{abs} of reaction (2) is related to the free energy change ΔG by Van't Hoff's equation:

$$-RT \ln K_{abs} = \Delta G(T) = \Delta H_0 + \int_0^T \Delta C_p \, dT - T \int_0^T \frac{\Delta C_p}{T} \, dT \quad (11)$$

where ΔH_0 is the enthalpy change at absolute zero and ΔC_p is the change in molar specific heat. Calculations on a molecular level always refer to the energy of an isolated molecule at absolute zero and thus give the quantity ΔH_0. Knowing the temperature dependence of ΔC_p one could thus calculate $\Delta G(T)$ and hence K_{abs}. Unfortunately, the exact temperature dependence of ΔC_p is seldom known, so an approximation has to be made. Usually it has been assumed [26, 27] that ΔH_0 is a satisfactory approximation to ΔG at finite temperatures.

$$\Delta G(T) \approx \Delta H_0 \quad (12)$$

Apparently the last two terms in eqn (11) cancel to a certain degree[28].

ΔH_0 for reaction (2) is the difference in heat of formation between R⁻, H⁺ and RH. Taking the energies of the isolated atoms as the energy zero, ΔH_0 can be equated with the difference in total bond energy (also called total energy of dissociation or atomization) E_b:

$$\Delta H_0 = E_b(R^-) + E_b(H^+) - E_b(RH) \quad (13)$$

In this equation the sign convention is such that $E_b(R^-)$ and $E_b(RH)$ are negative. $E_b(H^+)$ is identical with the ionization energy of hydrogen (314 kcal/mole). The total bond energies could be calculated using, e.g. POPLE's or DEWAR's semiempirical SCF MO methods[29a] which take into account all valence electrons and internuclear repulsions. However in the past it has been assumed that the total bond energy of a π-electron molecule can be divided into the total σ-bond energy $E_{\sigma b}$ and the total π-bond energy $E_{\pi b}$:

$$E_b = E_{\sigma b} + E_{\pi b} \quad (14)$$

In a recent series of papers DEWAR *et al.*[29, 31] have shown that $E_{\sigma b}$ can be calculated quite accurately from a refined set of bond increments while E_{nb} may be evaluated according to one of the current π-electron methods. Any non-bonded interactions[30] are disregarded in this approach. If it is assumed, as has frequently been done[32, 33], that for the dissociation of hydrocarbon acids the change in σ-bond energy is always the same (this is also likely to be true for the non-bonded interactions), one can absorb $E_{\sigma b}$ together with $E_b(\mathrm{H}^+)$ into a constant. Consequently, combining eqns (11) to (14), one obtains:

$$pK_{\mathrm{abs}} = a[E_{nb}(\mathrm{R}^-) - E_{nb}(\mathrm{RH})] + b \qquad (15)$$

where a is a constant depending only on the temperature and b is a dimensionless constant. Under the conditions just specified the acidity should thus be determined by the difference in π-bond energy between the acid and its anion. Simple Hückel theory (HMO theory) has been used in the majority of cases to evaluate this difference in π-bond energy. In the HMO theory, due to the neglect of electron repulsion, the π-bond energy E_{nb} of a π-system is identical with its π-energy E_π[29]. Expressing the π-energies in terms of the adjustable energy parameter β

$$E_\pi = M\beta \qquad (16)$$

(M being a dimensionless number obtained from the HMO calculation) one eventually arrives at the expression:

$$pK_{\mathrm{abs}} = [M(\mathrm{R}^-) - M(\mathrm{RH})]a\beta + b = \Delta Mc + b \qquad (17)$$

In this equation $b > 0$ and $c < 0$ are dimensionless empirical constants. Thus an acid is expected to be the more acidic the greater the ΔM-value is.

HMO Energies and Intrinsic Acidity

In *Table I*, ΔM-values are given for a number of hydrocarbons. In

Table I. ΔM-values for Hydrocarbons

No.[a]	Compound[b]	ϵ_m	$\dfrac{\epsilon_{m+1} - \epsilon_m}{[\epsilon_m - \epsilon_{m-1}]}$	ΔM	Ref.[c]
	Cyclopropene	−1·00	—	0	34
25	Cyclopentadiene	0·62	2·24	2·00	34
55	Cycloheptatriene	−0·45	1·36	1·11	34
26	Cyclononatetraene	0·35	1·35	2·00	34
53	Toluene	0	1·00	0·72	34
52	Diphenylmethane	0	1·00	1·30	34
			[1·00]		

Table I (continued)

No.[a]	Compound[b]	ϵ_m	$\dfrac{\epsilon_{m+1}-\epsilon_m}{[\epsilon_m-\epsilon_{m-1}]}$	ΔM	Ref.[c]
50	Triphenylmethane	0	1·00 [1·00]	1·80	34
35	Indene	0·30	1·20	1·75	34
43	Fluorene	0·18	0·99 [0·52]	1·52	34
66	1,3-Diphenylindene	0·20	0·98	2·21	34
39	1,2-Benzofluorene	0·20	0·94	1·62	35
41	2,3-Benzofluorene	0·11	0·68	1·49	35
38	3,4-Benzofluorene	0·22	0·89	1·58	35
33	1,2,7,8-Dibenzofluorene	0·23	0·90	1·69	35
42	2,3,6,7-Dibenzofluorene	0·07	0·60	1·45	35
30	3,4,5,6-Dibenzofluorene	0·28	0·92	1·63	35
40	4,5-Methylenephenanthrene	0·13	0·73	1·51	35
54	Propene	0	1·41	0·83	34
	1,3-Pentadiene	0	1·00	0·99	34
	1,4-Pentadiene	0	1·00	1·46	34
	1,3,5-Heptatriene	0	0·77	1·07	34
	1,3,6-Heptatriene	0	0·77	1·58	34
28a	BiphCH—CH=CH—Ph	0·11	0·79	2·08	35
28b	BiphC=CH—CH$_2$—Ph	0·11	0·79	1·66	35
31a	BiphCH—CH=CPh$_2$	0·09	0·72	2·15	35
31b	BiphC=CH—CHPh$_2$	0·09	0·72	2·12	35
23	BiphC=CH—CHBiph	0·18	0·73 [0·52]	2·23	36
67		0·30	0·74		36
18a	BiphC=CH—CH$_2$—CH=CBiph	0·15	0·58 [0·49]	1·93	36
18b	BiphC=CH—CH=CH—CHBiph	0·15	0·58	2·23	36
68a	BiphC=CH—CH=CH—CH$_2$—CH= CBiph	0·13	0·49 [0·42]	1·93	36
68b	BiphC=CH—(CH=CH)$_2$—CHBiph	0·13	0·49	2·23	36
11a	(BiphC=CH)$_3$CH	0·16	0·59 [0·48]	2·50	35
11b	(BiphC=CH)$_2$C=CH—CHBiph	0·16	0·59	2·34	35
27a	1,2,7,8-BiphCH—CH=CBiph	0·21	0·68	2·33	
27b	1,2,7,8-BiphC=CH—CHBiph	0·21	0·68	2·28	
16a	3,4,5,6-BiphCH—CH=CBiph	0·23	0·66	2·29	35
16b	3,4,5,6-BiphC=CH—CHBiph	0·23	0·66	2·27	35
10	3,4,5,6-BiphC=CH—CHBiph-3,4,5,6	0·28	0·64	2·33	35
9a	(3,4,5,6-BiphC=CH)$_2$CH$_2$	0·23	0·53	2·01	35
9b	3,4,5,6-BiphC=CH—CH=CH— CHBiph-3,4,5,6	0·23	0·53	2·31	35

Table I (continued)

No.[a]	Compound [b]	ϵ_m	$\dfrac{\epsilon_{m+1} - \epsilon_m}{[\epsilon_m - \epsilon_{m-1}]}$	ΔM	Ref.[c]
20a	PhenC=CH—CHBiph	0·15	0·71	2·21	35
20b	PhenCH—CH=CBiph	0·15	0·71	2·21	35
19	PhenCH—CH=CPhen	0·13	0·69	2·21	35
34	9-Phenylfluorene	0·14	0·93	1·98	34
32	9-Phenyl-1,2,7,8-dibenzofluorene	0·19	0·85	2·11	35
29	9-Phenyl-3,4,5,6-dibenzofluorene	0·22	0·82	2·06	35
22	BiphCH—C(Ph)=CBiph	0·18	0·62 [0·44]	2·18	35
24a	BiphC=CH—CH(Ph)—CH=CBiph	0·14	0·57	2·34	35
24b	BiphC=CH—C(Ph)=CH—CHBiph	0·14	0·57	2·26	35
21	Fluoradene	0·28	0·94	2·12	34
13	BiphC=CH—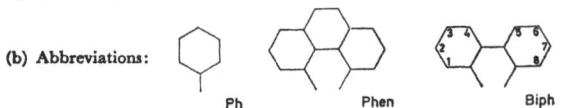=CH—CHBiph	0·22	0·58	2·35	35
	BiphC=CH⸺CH=CBiph BiphC=CH⟍CH=CBiph H CH=CBiph	0·29	0·48	2·51	35
17	9-Cyanofluorene	0·33	1·06	2·17	35[d]
		0·44	1·09	2·34	35[e]
14	Malononitrile	0·54	1·15	1·82	35[d]
		0·90	1·46	2·14	35[e]
6	Tricyanomethane	0·72	1·52	2·39	35[d]
		1·10	1·66	2·72	35[e]
5	1,1,3,3-Tetracyanopropene	0·54	1·01	2·48	35[d]
		0·90	1·20	2·80	35[e]
4	1,1,2,3,3-Pentacyanopropene	0·54	0·69	2·52	35[d]
		0·90	0·77	2·79	35[e]
3	1,1,3,3-Tetracyano-2-dicyanomethyl-propene	0·54	1·01	2·48	35[d]
	1,1,3,3-Tetracyano-2-dicyanomethyl-propenide	0·54	1·08	2·38	35[d]
	1,1,5,5-Tetracyano-1,4-pentadiene	0·43	0·76	2·11	35[d]
	1,1,5,5-Tetracyano-1,3-pentadiene	0·43	0·76	2·45	35[d]
7	1,1,2,6,7,7-Hexacyanoheptatriene	0·37	0·37	2·22	35[d]
		0·23	0·54	2·49	35[e]
1	Pentacyanocyclopentadiene	0·75	1·06	2·66	35[d]
		1·02	1·08	2·89	35[e]

(a) Numbering as in *Table X*

(b) Abbreviations: Ph   Phen   Biph

(Numbers added to the symbol 'Biph' indicate position of benzo-annelation)
(c) References to calculations.
(d) Heteroatom-parameters: $h_N = 1$, $k_{CN} = 1·2^{37}$.
(e) Heteroatom-parameters: $h_N = 2$, $k_{CN} = 1·2$.

addition, there may be found the energy of the highest occupied molecular orbital ϵ_m and the theoretical excitation energy $\epsilon_{m+1} - \epsilon_m$ of the corresponding anion. The method of calculation is simple Hückel theory neglecting any deviations from planarity and any bond length alternations.

Assuming the validity of eqn (17) several conclusions as to the intrinsic acidity of hydrocarbons can be drawn from this table:

(i) Cyclopropene and cycloheptatriene should be much less acidic than cyclopentadiene and cyclononatetraene.

(ii) In the series of completely planar cyclic polyenes $C_{4n+5}H_{4n+6}$ the acidity is expected to be independent of ring size.

(iii) Benzo-annelation on to cyclopentadiene should decrease the acidity, while further benzo-annelation on to fluorene in the 1,2- or 3,4-position should increase the acidity again.

(iv) In the series of linear polyenes $C_{2n+1}H_{2n+4}$ the acidity is expected to increase steadily with increasing chain length. However, upon introduction of substituents in α,ω-position a decrease in acid strength may even result with increasing chain length.

(v) The acid strengths of tautomeric polyenes should be markedly different.

(vi) Simple cyanocarbon acids should exhibit a considerable acidity.

HMO Energies and Experimental Acid Strength

A comparison between theory and experiment cannot be based on eqn (17) since K_{abs}-values are not accessible to measurement. The measurable thermodynamic dissociation constant $_sK$ differs from K_{abs} by a term relating to the change in free energy of solvation:

$$p_sK = pK_{abs} + a\Delta G_{solv} \tag{18}$$

Thus it appears necessary to calculate ΔG_{solv} in order to get a relationship between measured p_sK-values and calculated energy quantities. The exact calculation of ΔG_{solv} is a problem unsolved as yet, although part of the various contributions to the solvation energy can be estimated to some extent[30, 38].

In solvents of low hydrogen bonding ability normally employed in the field of acidic hydrocarbons, an important contribution to the solvation energy comes from an isotropic charge–dipole or dipole–dipole interaction between solvent and solute. BORN[39] has put forward a formula, recently modified by HOIJTINK *et al.*[40] which allows the estimation of these interactions:

$$E_{\text{solv}} = -\sum_{i,j} \frac{q_i q_j}{2r_{ij}} \left(1 - \frac{1}{D}\right) \qquad (19)$$

where D is the dielectric constant of the solvent, q_i is the net charge on atom i of the solute and r_{ij} is the distance between atom i and j. For $i=j$, r_{ii} represents an effective radius of atom i.

It is apparent from this certainly very crude formula that E_{solv} becomes numerically the smaller the more the charge is distributed over the molecule. If ΔE_{solv} calculated according to eqn (19) were a satisfactory approximation to ΔG_{solv} at room temperature a dependence of pK on the size of the molecule would be the consequence. It would therefore be highly desirable to evaluate at least that part of the solvation energy which is represented by formula (19) if molecular energies are to be compared with acid strengths as given by $_sK$. This was done by CHALVET *et al.*[30] for the equilibrium $BH^+ \rightleftharpoons B + H^+$ (B = pyridine, chinoline, acridine): The change in π-energy increases in this series but this is nearly compensated for by a decrease in solvation energy change in agreement with the experimental finding that the p_sK values are approximately equal.

In the field of acidic hydrocarbons solvation energies according to formula (19) have apparently never been calculated although JANO[41] has recently recast this formula in a form especially suitable for π-electron calculations. Thus the comparison of molecular energies with p_sK-values has so far been based almost exclusively on the assumption that ΔG_{solv} is constant. We may now give a precise meaning to the term 'similar behaviour' introduced in the previous section (p. 122): A series of acids is said to behave similarly if in a particular solvent ΔG_{solv} is approximately the same for all acids. Combining eqns (17) and (18) and absorbing $a\Delta G_{\text{solv}}$ into the constant b, one obtains[32]

$$p_sK = \Delta Mc + b' \qquad (20)$$

In this eqn b' and c are empirical constants and ΔM is a number to be calculated from HMO theory. On the basis of eqn (20) the ΔM-values recorded in *Table I* are therefore also relevant for a comparison with experimental acid strength.

It seems appropriate to summarize briefly the major approximations made in deriving eqn (20): (i) $\Delta G(T)$ is equated with ΔH_0; (ii) ΔG_{solv} is set constant for a series of structurally related acids measured in the same solvent; (iii) the change in σ-bond energy is neglected or treated as a constant; (iv) non-bonding interactions (steric effects) are neglected; (v) the change in π-bond energy is equated with the change in π-energy calculated according to simple Hückel theory.

MEASUREMENT OF CH-ACIDITY

According to the arguments presented in foregoing sections, the most reliable thermodynamic acidity scales should be those measured for a series of structurally similar acids in a solvent of high dielectric constant. In principle, any such solvent could be used. However, it has become customary to refer acid strengths to water, i.e. the pK-value of an acid is understood to be the negative logarithm of the thermodynamic dissociation constant $_wK$ of eqn (4) measured in water. This convention raises problems for hydrocarbons which are usually insoluble in water. In addition, acidic hydrocarbons are in general such weak acids, that there would not be any detectable dissociation in water. Therefore, the determination of p_wK-values of hydrocarbons involves two steps: (i) measurement of K_r-values in an appropriate solvent by equilibrium or kinetic methods; (ii) transfer of these K_r-values to water.

Measurement of K_r-values by Equilibrium Methods

These methods are based on equilibria corresponding to eqn (5). The concentrations of the carbanions can in most cases be easily determined by optical methods since the anions are in general coloured and absorb at considerably longer wavelengths than the hydrocarbons. B may represent the anion of a second hydrocarbon or some other base.

Equilibria Involving a Second Acidic Hydrocarbon—The equilibrium constant K_r of the reaction $R_1H + R_2^- \rightleftharpoons R_1^- + R_2H$

$$K_r = \frac{[R_1^-] \cdot [R_2H]}{[R_2^-] \cdot [R_1H]} \tag{21}$$

is equal to the ratio of the $_sK$-values of R_1H and R_2H (eqn (6)). The position of the equilibrium can either be estimated by carbonation and isolation of the resulting carboxylic acids[11] (assuming that the rate of carbonation surpasses the rate for attainment of the equilibrium), or more directly by optical methods. The method based on eqn (21) was originally introduced by Conant and Wheland[11] and by McEwen[12]

128

and has been used more recently by STREITWIESER *et al.*[42] and in some cases by STEINER and GILBERT[43]. Some of their results are compared in *Table II*.

Table II. Comparison of pK_r-values for Hydrocarbons, Obtained in Various Solvents by Equilibrium Methods

Hydrocarbon	pK_r			
	Ether or benzene[12]	Cyclohexyl-amine[42]		Dimethyl-sulphoxide[43]
		Li-salts	Cs-salts	
Diphenylmethane	+2·5	+1·6	+1·6	+1·4
Triphenylmethane	0	0	0	0
p-Diphenylyldiphenyl-methane	−1·5	−1·3	−1·3	−1·9
2,3-Benzofluorene	—	−9·3	−8·3	−9·0
Fluorene	−7·5	−9·6	−8·8	−8·3
4,5-Methylenephenanthrene	—	−9·9	−8·9	−9·0
Indene	−11·5	−12·2	−11·6	−10·6
9-Phenylfluorene	11·5	−14·0	−13·0	−12·2

As can be seen, there is good agreement, though in ether and cyclo-hexylamine ($D = 4·5$ and 5 respectively) ion pair formation will predominate and the equilibrium constant actually measured is a ratio of two ionization constants.

Recent investigations demonstrate that metal compounds of hydro-carbons are associated to a considerable extent in solvents of low dielectric constant. As early as 1922, ZIEGLER and WOLLSCHITT[44], on the basis of conductivity measurements, postulated ion pair association in ether. Conductivity measurements in cyclohexylamine[45] pointed to ion pair dissociation constants as low as 10^{-10} for lithium cyclohexyl-amide, lithium perchlorate and fluorenyl-lithium. In tetrahydrofuran, dimethoxyethane and diethyl ether, fluorenyl-lithium seems to exist as a well-defined solvent complex[46]. HOGEN-ESCH and SMID[47] interpret ultra-violet spectral data of solutions of fluorenyl-sodium in tetrahydro-furan in the sense that there are two different kinds of ion pairs (contact and solvent separated ion pairs)[48] the ratio of which depends on temperature. In contrast to this work HOIJTINK *et al.*[49] concluded from conductivity measurements that lithium salts of certain aromatic hydro-carbons in tetrahydrofuran are completely dissociated. Hoijtink's results, however, are at variance with those of a recent investigation by SLATES and SWARC[50] who obtained ion pair dissociation constants of about 10^{-5} for the corresponding sodium compounds. Highly dissociated salts of hydrocarbons seem to exist in dimethylsulphoxide (DMSO) according to preliminary conductivity measurements[43]. In agreement with this contention is the fact that the ultra-violet spectra of fluorenyl-sodium and fluorenyl-potassium in DMSO[43, 51] are practically identical with those of Hogen–Esch's and Smid's solvent separated ion pairs.

Equilibria Involving a Suitable Base—Dimethylsulphinylanion[43], alkoxide[43, 52], hydroxide[43, 53], hydrazine[54], trialkylamines[51, 55] and acetate[51] have been used as base B in equilibrium (5) depending on the strength of the acid RH. The relative acid strength of two hydrocarbons (R_1H, R_2H) is obtained as the ratio of the corresponding equilibrium constants (eqn (6)). The trouble with this type of experiment is that only a relatively narrow range of p_sK-values can be covered with a particular solvent/base system. In order to obtain a relative p_sK-scale of acids of very different strength several different solvent/base systems have to be employed. Two such solvent/base systems may be connected by measuring the equilibrium constant for suitable hydrocarbons in both systems. However, it must be borne in mind that changing the solvent introduces the uncertainties as to the solvation energies discussed in detail in the previous two sections. Therefore this method is limited to structurally similar acids (for which the change in free energy of solvation may be assumed to be the same).

A particular useful solvent/base system is a mixture of DMSO with alcohols, water or acetic acid and alkoxide, hydroxide or acetate as base. The basicity of alkoxide[56] or hydroxide[53] is enormously increased with increasing DMSO content of the solvent, especially above 95 mole per cent DMSO[57]. Similarly enhanced is the *kinetic activity* of bases in DMSO. Base-catalysed isomerization[58], racemization[59], and hydrogen–isotope exchange[60] reactions in DMSO proceed by a factor of about 10^{10} faster than in alcohols.

CH-acidity measurements in a fixed solvent/base system were first carried out by STEARNS and WHELAND[61] using ethanol/ethoxide; tris-(4-nitrophenyl)-methane and some NH-acids were investigated. Steiner and Gilbert[43] used the system DMSO/dimsylpotassium and determined relative pK-values for DMSO, triphenylmethane and diphenylmethane. KUHN and REWICKI[51] measured relative acid strengths of some of the more acidic hydrocarbons in DMSO/10^{-3} m tripropylamine. Their results are shown in *Table III*.

Acidity measurements on a series of acids using more than a single

Table III. pK_r-values of Acidic Hydrocarbons in $DMSO/10^{-3}$ m Tripropylamene[51]; the pK-value of 9-cyanofluorene was taken as zero

Hydrocarbon[a]	pK_r
3,4,5,6-BiphCH—CH=CBiph	−0·3
9-Cyanofluorene	0
BiphCH—CH=CH—CH=CBiph	0·9[b]
PhenCH—CH=CPhen	2·1
Fluoradene	2·4
BiphCH—C(Ph)=CBiph	2·5
BiphCH—CH=CPhen	2·6[b]
BiphCH—CH=CBiph	2·8
BiphCH—CH=C(Ph)—CH=CBiph	3·4[b]

(a) Abbreviations, cf. Tables I or X.
(b) Gross acidity (p. 134).

solvent/base system are reviewed in *Table IV*. The more important results of these measurements are collected in *Tables V–VIII*.

Table IV. Review of Solvent/Base Systems used for Relative Acidity Measurements According to eqn (5)

Solvent system	Base	Compounds measured	Ref.
$N_2H_4/H_2O(35-100\%)$	—	NH-acids, bis- and tris-[4-nitrophenyl]-methane	54
Sulfolane/$H_2O(3-35\%)$	PhN(CH$_3$)$_3$]OH	NH-acids, 9-Phenyl-fluorene	53
DMSO/$C_2H_5OH(5-100\%)$	C_2H_5ONa	Bis- and tris-[4-nitro-phenyl]-methane, fluorene derivatives	51, 52
DMSO/$H_2O(0-6\%)$	KOH	Arylmethanes, fluorene derivatives	43
DMSO/$CH_3OH(0-7\%)$	CH_3OK	Arylmethanes, fluorene derivatives	43
DMSO/$CH_3CO_2H(0-30\%)$	CH_3CO_2Na	Fluorene derivatives	51
$H_2SO_4/H_2O(1-70\%)$	—	Cyanocarbon acids	62
$HClO_4/H_2O(1-70\%)$	—	Cyanocarbon acids	62

In the case of tautomeric hydrocarbons (R_1H, R_2H) the following equilibria are encountered:

$$R_1H + B \overset{K_T}{\rightleftharpoons} R_2H + B$$
$$K_1 \searrow \qquad \nearrow K_2 \tag{22}$$
$$R^- + BH^+$$

131

Table V. pK_r-values of Weak CH-acids in Different Solvent/Base Systems According to STEINER *and* GILBERT[43]; *the pK-value of triphenylmethane was taken as zero*

Compound	pK_r [a]		
	DMSO/H$_2$O	DMSO/CH$_3$OH	DMSO
Dimethysulphoxide	—	—	4·1
Diphenylmethane	—	—	1·4
9-Methylanthracene	—	0·5	—
Triphenylmethane	0	0	0
Xanthene	—	−0·1	—
Diphenylyldiphenylmethane	—	−1·9	—
9-Phenylxanthene	−2·9	−3·0	−3·0
Tris-(diphenylyl)methane	—	−4·4	—
Fluorene	−6·7	−6·7	−6·7
4-Nitroaniline	−8·8	−8·7	−8·6
Indene	—	−9	—
2,4-Dinitroaniline	−12·2	−12·7	−12·5
2,4-Dinitrodiphenylamine	−13·4	−14·0	—

(a) In a recent communication STEINER and STARKEY[43] point to an error in the earlier paper. However no detailed corrections were given for the various solvent systems.

Table VI. pK_r-values of Acidic Hydrocarbons in DMSO/Ethanol/Sodium Methoxide According to KUHN *and* REWICKI[51]; *the pK-value of 9-phenylfluorene was taken as zero*

Hydrocarbon[a]	pK_r
2,3,6,7-Dibenzofluorene	+2·9
9-Phenylfluorene	0
1,2,7,8-Dibenzofluorene	−1
BiphC=CH—CH(Ph)$_2$	−1 [b]
9-Phenyl-1,2,7,8-dibenzofluorene	−1·2
BiphC=CH—CH$_2$—Ph	−1·6 [b]
3,4,5,6-Dibenzofluorene	−1·7
9-Phenyl-3,4,5,6-dibenzofluorene	−3·7
BiphC=CH—CHBiph-1,2,7,8	~ −3·5

(a) Abbreviations, cf. *Tables I* or *X*.
(b) Gross acidity (p. 134).

By measuring the equilibrium concentration of R_1H and R_2H one obtains the ratio of the two thermodynamic dissociation constants

$$K_T = \frac{[R_2H]}{[R_1H]} = \frac{K_1}{K_2} \tag{23}$$

The gross dissociation constant K_g of the pair of tautomers is defined by

$$K_g = \frac{[R^-].[BH^+]}{[R_1H]+[R_2H]} = \frac{K_1K_2}{K_1+K_2} \tag{24}$$

Table VII. pK_r-values of Acidic Hydrocarbons in DMSO/Acetic Acid/10^{-2} m Sodium Acetate According to KUHN *and* REWICKI[51]; *the pK-value of 9-cyanofluorene was taken as zero*

Hydrocarbon[a]	pK_r
BiphC=CH—CHBiph-1,2,7,8	~4·5
BiphC=CH—CH(Ph)—CH=CBiph	3·5[b]
BiphC=CH—CHBiph	3·0
BiphC=C(Ph)—CHBiph	2·7
Fluoradene	2·6
PhenC=CH—CHPhen	2·3
BiphC=CH—CH=CH—CHBiph	0·9[b]
9-Cyanofluorene	0
BiphC=CH—CHBiph-3,4,5,6	−0·2[b]
[BiphC=CH]$_2$C=CH—CH=CH—CHBiph	−0·2[b]
[BiphC=CH]$_3$CH	−1·0[b]
BiphC=CH— 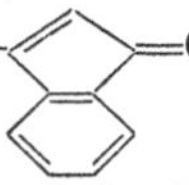 =CH—CHBiph	−1·4[b]
3,4,5,6-BiphC=CH—CH=CH—CHBiph	−1·6[b]
3,4,5,6-BiphC=CH—CHBiph-3,4,5,6	−2·3[b]
3,4,5,6-BiphC=CH—CH=CH—CHBiph-3,4,5,6	−3·2[b]

(a) Abbreviations, *cf. Tables I* or *X.*
(b) Gross acidity (p. 134).

Table VIII. pK_r-values of Cyanocarbon Acids in H$_2$SO$_4$/H$_2$O or HClO$_4$/H$_2$O According to BOYD[62]; *the pK-value of p-(tricyanovinyl)phenyl-dicyanomethane was taken as zero*

Compound	pK_r	
	H$_2$SO$_4$/H$_2$O	HClO$_4$/H$_2$O
p-(Tricyanovinyl)phenyl-dicyano-methane	0	0
Methyldicyanoacetate	−1·3	−1·2
1,1,2,6,7,7-Hexacyanoheptatriene	−3·3	−3·0
Tricyanomethane	−4·4	−4·5
Tricyanovinyl alcohol	−4·4	−4·7
Bis-(tricyanovinyl)amine	−5·4	−5·5
Tetracyanopropene	—	< −8
Pentacyanopropene	—	< −8·5
Hexacyanoisobutene	—	< −8·5(pK_1)

If K_T and K_g are measured separately it is possible to calculate the individual dissociation constants K_1 and K_2. This has been accomplished in a few cases[55].

Another method for the determination of individual pK-values of tautomers is based on the assumption that the dissociation equilibrium

is set up much more rapidly than the equilibrium between the tautomers. Thus extrapolating the degree of dissociation of one tautomer to
zero time one may derive the pK-value of that tautomer. For the hydrocarbons investigated [55] both methods led to nearly identical pK-values.
Table IX lists pK_T-values for some pairs of tautomeric hydrocarbon
acids[51]. Using the gross dissociation constants as given in *Tables IV* and
VI one may calculate the individual pK-values for each of the tautomers (*cf. Table X*).

Table IX. pK_T-values of Pairs of Tautomeric Hydrocarbon Acids in DMSO[51]

Tautomers		$-pK_T$
BiphCH—CH=CH—Ph	⇌ BiphC=CH—CH₂—Ph	> 2
BiphCH—CH=CPh₂	⇌ BiphC=CH—CHPh₂	0·45
BiphC=CH—CHPhen	⇌ BiphCH—CH=CPhen	0·45
BiphC=CH—CH₂—CH=CBiph	⇌ BiphC=CH—CH=CH—CHBiph	0·40
(BiphC=CH)₃CH	⇌ (BiphC=CH)₂C=CH—CHBiph	0·95

Further Equilibrium Methods—The following equilibria

$$RLi + R'I \rightleftharpoons RI + R'Li$$
$$R_2Mg + R_2'Hg \rightleftharpoons R_2Hg + R_2'Mg$$
$$R = \text{alkyl, aryl, alkenyl}$$

have been used to estimate relative stabilities of the anions of these
extremely weak hydrocarbon acids[63, 64]. The results of the two methods
agree well; RAPOPORT and SMOLINSKY[65] measured the p_sK-value of
fluoradene in 97% methanol from the distribution equilibrium between
heptane and buffer solutions in 97% methanol. Recently RITCHIE and
USCHOLD[69a] determined p_sK-values of several hydrocarbons in DMSO
by potentiometric titration using a glass electrode.

Measurement of K_r-values by Kinetic Methods

Kinetic methods, especially hydrogen isotope exchange methods, have
usually been applied in order to determine *kinetic acidities* of hydrocarbons. The extensive work in this field is summarized elsewhere[66].
Only the attempts to connect kinetic with equilibrium acidity by means
of the Brönsted relation (1) will be mentioned here. Rate constants of
protolytic reactions involving various CH-acids (but no hydrocarbons)
have been measured and correlated with the corresponding p_wK-
values[67, 68]. SHATENSHTEIN *et al.*[69] related the velocity of the H/D
exchange for diphenylmethane, triphenylmethane, fluorene and indene
in NH₃/KNH₂ with McEwen's pK-values. Dessy *et al.*[25] performed the

same type of experiment in dimethylformamide/triethylamine with cyclopentadiene, phenylacetylene, fluorene, acetylene and various substituted acetylenes using also McEwen's pK-values. Streitwieser Jr. et al.[42] measured relative pK-values of a number of fluorene derivatives in cyclohexylamine and correlated these with the rate constants for the H/T exchange in methanol/sodium methylate. Relatively good Brönsted plots are found in all three cases mentioned with $\alpha \approx 0.4$, but in no case have these plots been used to establish unknown pK-values. Thus, on the whole, kinetic methods have so far been only of minor importance for the determination of equilibrium acidities. The abundant rate data available for the hydrogen isotope exchange for simple aliphatic and aromatic hydrocarbons can at present be used only for a rather crude estimate of the corresponding pK-values. Thus CRAM[78] assumed that an increase in rate of exchange by one power of ten corresponds to a decrease of pK by about one unit.

pK-values of Acidic Hydrocarbons Referred to Water

For comparison with other acids it is intended now to refer all the relative acid strengths of hydrocarbons to water as the common solvent. However, it should be stated at once that this is a problem which has not yet received a final answer. This is exemplified by the p_wK-values of fluorene to be found in the literature: 20.5^{43}, 21^{52}, 22.9^{42}, 25^{12}, 31^{70}. These deviations may in part come from errors in measuring relative acid strengths, but the main deficiency is undoubtedly due to the approximation which is nearly always made that relative acid strengths are independent of the solvent system. As repeatedly emphasized in previous sections, this may be a satisfactory approximation in cases of similar acids. However, a p_wK-value of a hydrocarbon can as yet not be directly determined in water, therefore reference to some standard acid of known p_wK-value has to be made. The standard acids are not in general similar to hydrocarbons.

If the relative acid strengths of the hydrocarbons and the standard acid were independent of solvent the transformation of the relative pK-values to p_wK-values would involve two steps: (i) determination of the p_wK of a standard acid; (ii) determination of the acid strength of the standard acid relative to any hydrocarbon.

The standard acids suitable for this purpose are usually much too weak to be measured directly in water consequently their p_wK-values have to be established indirectly. This can conveniently be done by a stepwise procedure[71] involving an H_- acidity function[72, 73]. Ideally

135

H$_-$ is a function of the solvent composition only and indicates the power of a solvent to remove a proton from a neutral acid. H$_-$ is defined in analogy to the pH-value by

$$H_- = p_wK + \log \frac{[R^-]}{[RH]} = -\log a_{H^+} \cdot \frac{f^*_{R^-}}{f^*_{RH}} \quad (25)$$

where the f^*'s are activity coefficients normalized to unity in ideal aqueous solution. The H$_-$ scale of a solvent/base system can be established by measuring the ratio $[R^-]/[RH]$ of a series of acids of known p_wK-values (indicators). Knowing the H$_-$ value, the p_wK of any acid RH is obtained by measuring the ratio $[R^-]/[RH]$ at a suitable solvent composition.

In fact, for acids of different structure the ratio $f^*_{R^-}/f^*_{RH}$ (eqn 25) cannot be assumed to be the same at a given solvent composition. Thus H$_-$ scales depend on the types of indicators used[74]. The H$_-$-technique gives good results if the acids are of the same type as the indicator acids. For example, the p_wK-values of a number of weak NH-acids obtained in *different* solvent systems and based on *different* standard acids agree quite well[75] indicating that such p_wK-values are meaningful.

In the field of acidic hydrocarbons, however, there are only a few independent attempts to construct p_wK-scales. On the whole these scales agree roughly, especially for some of the more acidic hydrocarbons. Making this statement we omit a p_wK-scale established by STREITWIESER JR. *et al.*[70] and CRAM[76] since it was mainly based on theoretical considerations rather than experimental measurements. Moreover STREITWIESER JR. *et al.*[42, 77] and also Cram[78] have apparently given up this scale meanwhile.

McEwen[12] has put forward the first p_wK-scale of acidic hydrocarbons using semiquantitative colorimetric, spectroscopic and polarimetric methods. He based his scale on the standard acid methanol, the p_wK-value of which was taken as 16. Methanol was then related to other alcohols and 9-phenylfluorene in benzene. The acid strength of further hydrocarbons was referred to 9-phenylfluorene. Three points of criticism may be raised against this frequently cited scale: (i) in benzene strong ion association will prevail therefore the limitations discussed on p. 122 will apply; (ii) methanol or other alcohols cannot be expected to behave similarly to 9-phenylfluorene; (iii) McEwen found $\Delta pK = 2$ between methanol and ethanol, while according to more recent measurements, this difference is only about 0·5[79]. If McEwen's scale is reconstructed starting with methanol ($p_wK = 15·5$[79]), using his more quantitative

136

polarimetric measurements, the p_wK-value of 9-phenylfluorene (and therefore those of the other hydrocarbons) is reduced by about 2 units. The p_wK-scale thus obtained would be in much better agreement with more recent scales.

LANGFORD and BURWELL[53] as well as Steiner and Gilbert[43] used 4-nitroaniline ($p_wK = 18\cdot4$) as standard acid and so did Streitwieser Jr. et al.[42] since they based their measurements on Langford and Burwell's p_wK-value for 9-phenylfluorene. The p_wK-value of 4-nitroaniline was obtained from various H_-scales based on NH-acids[75]. However, it is questionable whether 4-nitroaniline and 9-phenylfluorene can be assumed to behave similarly. Indeed Steiner and Starkey[43] found that the pK-difference between hydrocarbons and nitroanilines in aqueous DMSO depends strongly on the solvent composition.

BOWDEN and STEWART[52] constructed an H_-scale for the system DMSO/EtONa using only CH-acids as indicators. They started with malononitrile in water and established the p_wK-values of 9-cyano-fluorene, 9-methoxycarbonylfluorene, tris- and bis-(4-nitrophenyl)-methane. The latter compound was compared with 9-phenylfluorene and this with fluorene. The p_wK-value of 9-phenylfluorene thus obtained is in good agreement with Langford and Burwell's value. However it is by no means certain that the various substituted CH-acids used are really more similar to hydrocarbon acids than NH-acids. In case of nitro- and cyano-substituted compounds, for example, BOYD[80] observed a different dependence of the corresponding mean activity coefficients on solvent composition. Moreover, the extinction co-efficients of the anions strongly depend on the solvent and the cation in case of the nitroarylmethanes[81]. Finally, it should be noted also that Bowden and Stewart's extinction coefficients for the anions of 9-phenyl-fluorene and fluorene in DMSO are at variance with those reported elsewhere[42, 43, 51].

Kuhn and Rewicki[51] based their measurements on Bowden and Stewart's p_wK-value of 9-cyanofluorene. A series of highly acidic hydro-carbons was then related to 9-cyanofluorene using the systems DMSO/tri-n-propylamine and DMSO/CH$_3$CO$_2$H/CH$_3$CO$_2$Na, respectively (*Tables III* and *VII*). In the overlapping area the p_wK-values agree well. The gap to 9-phenylfluorene was bridged by hydrocarbon 29 of *Table X*. Using the pK_r-values listed in *Table VI* a p_wK-value of about 19 results for 9-phenylfluorene in good agreement with the values obtained by other methods. On the other hand, fluoradene is found to have $p_wK = 14$ in disagreement with Rapoport and Smolinsky's[65] value

Table X. p_wK-values for Cyanocarbon Acids and Hydrocarbon Acids

No.	Compound[a]	Ref.[b]	Anion[c] λ$_{max}$[mμ]	Anion[c] Ref.	p_wK[d]	Ref.
1	Pentacyanocyclopentadiene	83	291 (CH$_3$CN)	152	< −10[e]	83
2	Tetracyanocyclopentadiene	83	298 (CH$_3$CN)	152	< −10*[e]	83
3	(NC)$_2$C=C[CH(CN)$_2$]$_2$	84	370 (H$_2$O)	62	< −8·5	62
4	1,1,2,3,3-Pentacyanopropene	84	395 (H$_2$O)	62	< −8·5	62
5	1,1,3,3-Tetracyanopropene	84	344 (H$_2$O)	62	< −8	62
6	Tricyanomethane	85	210 (H$_2$O)	62	−5·1	62
7	1,1,2,6,7,7-Hexacyanoheptatriene	84	645 (H$_2$O)	62	− 3·7*	62
8	p-(Tricyanovinyl)phenyldicyanomethane	86	607 (H$_2$O)	62	0·6[f]	62
9	(3,4,5,6-Biph C=CH)$_3$CH	148	697 (DMSO)	148	5·9*	148
10	(3,4,5,6-BiphC=CH)$_2$CH$_2$	87	694 (DMSO)	82	8·2*	51
11a	3,4,5,6-BiphC—CH—CHBiph-3,4,5,6	87	617 (DMSO)	82	9·1	51
	(BiphC=CH)$_3$CH	87–90	647 (DMSO) [~675]	82	9·4	51
12	3,4,5,6-BiphC=CH—CH$_2$—CH=CBiph	87	659 (DMSO)	82	9·8*	51
13	BiphC=CH 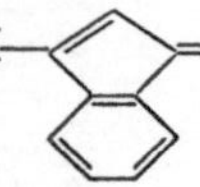 ⬡ =CH—CHBiph	87	694 (DMSO)	82	10·0*	51
11b	(BiphC=CH)$_2$C=CH—CH=CHBiph	89	647 (DMSO)	82	10·4	51
14	Malononitrile	52	225 (H$_2$O)	52	11·2	52
15	(BiphC=CH)$_2$C=CH—CH=CH—CHBiph	87	681 (DMSO) 652	82	11·2*	51
16	3,4,5,6-BiphCH—CH=CBiph	87	584 (DMSO)	82	11·2*	51
17	9-Cyanofluorene	91	445 (DMSO)	51	11·4[g]	52
18a	BiphC=CH—CH$_2$—CH=CHBiph	91	634 (DMSO) [678]	82	11·8	51
18b	BiphC=CH—CH—CH—CHBiph	92	634 (DMSO)	82	12·2	51

Table X (continued)

No.	Compound[a]	Ref.[b]	Anion[c] λmax[mμ]	Ref.	p_wK[d]		Ref.
19	PhenC=CH—CHPhen	93	558 (DMSO)	82	13·6		51
20b	PhenCH—CH=CBiph	93	558 (DMSO)	82	13·6		51
21	Fluoradene	65	564 (DMSO)	82	13·9		51
22	BiphC=C(Ph)—CHBiph	91	600 (DMSO) [773]	82	14·0		51
20a	BiphCH—CH=CPhen	93	558 (DMSO)	82	14·0		51
23	BiphCH—CH=CBiph	91	559 (DMSO) [655]	82	14·3		51
24	BiphC=CH—CH(Ph)—CH=CBiph	87, 88	684 (DMSO)	82	14·8*		51
25	Cyclopentadiene		220 (ether)	155	14–16[h]		25
26	Cyclononatetraene	94	322 (CH₃CN)	94	—	>14–16	94
27	1,2,7,8-BiphCH—CH=CBiph	87	611 (DMSO)	82	~16*		51
28a	BiphCH—CH—CH—Ph	51	530 (DMSO) 488	82	<15·5		51
29	9-Phenyl-3,4,5,6-dibenzofluorene	51	508 (DMSO)	82	15·9		51
30	3,4,5,6-Dibenzofluorene	95	443 (DMSO)	82	16·8		51
28b	BiphC=CH—CH₂—Ph	96	530 (DMSO) 488	82	16·9		51
31a	BiphCH—CH=CPh₂	97	540 (DMSO) 515	82	17·1		51
31b	BiphC=CH—CHPh₂	91	540 (DMSO) 515	82	17·5		51
32	9-Phenyl-1,2,7,8-dibenzofluorene	98	473 (DMSO)	82	17·3		51
33	1,2,7,8-Dibenzofluorene	99	467 (DMSO)	82	17·5		51
34	9-Phenylfluorene		521 (DMSO)	82	18·5[l] 18·6	18·5[l] 19	53 42 / 52 12[α]

139

Table X (continued)

No.	Compound(a)	Ref.(b)	Anion(c) λmax[mμ]	Ref.	pwK(a)	pwK(a)	Ref.
35	Indene		380 (DMSO)	43	20·1	19·9	43 42
36	9-α-Naphthylfluorene		—	—	—	19	12(α)
37	Phenylacetylene		—	—	—	19	12(α)
38	3,4-Benzofluorene		519 (CHA)	42	19·6	19·4	149 42
39	1,2-Benzofluorene		425 (CHA)	42	—	20	42
40	4,5-Methylenephenanthrene		540 (DMSO) 510	51	21·2	22·6	149 42
41	2,3-Benzofluorene	95	657 (CHA)	42	<21·4	23·2	43 42
42	2,3,6,7-Dibenzofluorene		491 (DMSO)	51	21·4	—	51
43	Fluorene		516 (DMSO) [655]	51 100	22·4 22·1	22·7 23	43 42 149 12(α)
44	(Ph—C₆H₄)₃CH		635 (DMSO)	43	26·3	—	43
45	Acetylene		—	—	—	~25(1)	101
46	1,1,3-Triphenylpropene		556 (CHA)	42	—	26·5*	42
47	9,9-Dimethyl-10,10-diphenyldihydroanthracene		454 (CHA)	42	—	29·0	42
48	(Ph—C₆H₄)₂CH—Ph		595 (CHA)	43	28·8	31·2	43 42
49	4-Benzyldiphenyl		548* (DMSO)	43	—	29	12(α)
50	Triphenylmethane		500 (DMSO) [424]	43	30·8 30·7	— 32·5, 31·5	43 43 42
51	9-Methylanthracene		470 (DMSO)	43	31·2	—	43
52	Diphenylmethane		460 (DMSO) [440]	43	32·1	33·1	43 42
53	Toluene (α-position)					35	78
54	Propene (α-position)					35·5	78
55	Cycloheptatriene					36	78

140

Table X (continued)

No.	Compound(a)	Ref.(b)	Anion(c)		p_wK(d)		Ref.
			$\lambda_{max}[m\mu]$	Ref.			
56	Ethylene				36·5		78
57	Benzene				37		78
58	Cumene (α-position)				37		78
59	Tripticene (α-position)				38		78
60	Cyclopropane				39		78
61	Methane				40		78
62	Ethane				42		78
63	Cyclobutene				43		78
64	Cyclopentane				44		78
65	Cyclohexane				45		78

(a) Abbreviations:

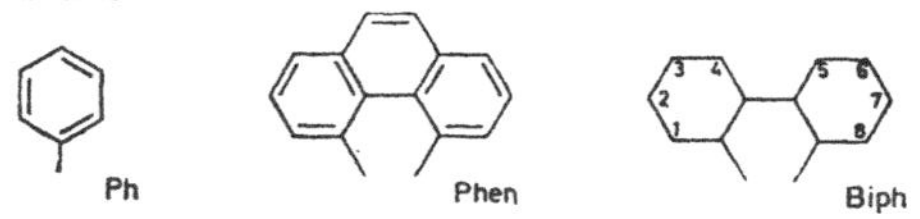

Ph

Phen

Biph

(Numbers added to the symbol 'Biph' indicate position of benzoannelation)

(b) References to syntheses.

(c) Abbreviations: DMSO = dimethysulphoxide; CHA = cyclohexylamine. Numbers in brackets [...] indicate λ_{max}-values of corresponding cations.

(d) p_wK-values marked by an asterisk denote gross p_wK-values. First column: p_wK-values based on measurements in solvents of high dielectric constant. Second column: p_wK-values based on measurements in non-polar solvents.

(e) Values estimated following Webster[87] who compared the acid strength with that of $HClO_4$ in acetonitrile[102], p_wK of $HClO_4$ was taken as -10^{103}.

(f) Standard for compounds 3–7.

(g) Standard for compounds 9–13, 15, 16, 18–24, 27, 29.

(h) No exact equilibrium measurements have as yet been reported; strong dissociation is claimed[104] in liquid ammonia, and water seems to be the weaker acid in this solvent[105]

(i) Standard for compounds 28,30–65; the values of Steiner *et al.*[43] were altered to conform with this standard.

(k) McEwen's scale was shifted to lower p_wK-values by two units (*cf.* p. 139).

(l) 9-Alkylfluorenes and alkylacetylenes seem to be of comparable acid strength too[106].

141

($p_wK = 11$). The distribution method of these authors has apparently never been applied to any other hydrocarbon besides fluoradene, and its limitations cannot at present be properly assessed. For example, it is uncertain how far salt effects will influence the distribution equilibrium of fluoradene and the standard acid 9-(o-hydroxyphenyl)-fluorene between heptane and buffer solutions in 97 per cent methanol.

The most reliable H_--scale is probably the one obtained in H_2SO_4/H_2O (or $HClO_4/H_2O$) by Boyd[62] using cyanocarbon acids as indicators. These acids are of similar structure and size and are in addition soluble in water. So the p_wK-values of the cyanocarbon acids seem to be the most trustworthy ones of *Table X*.

Table X represents an attempt to construct a consistent p_wK-scale for hydrocarbon acids and cyanocarbon acids. The longest wavelength transitions of the corresponding anions are also given.

INTERRELATION OF THEORETICAL AND EXPERIMENTAL RESULTS

Comparison of UV-spectra of Anions of Acidic Hydrocarbons with Calculations

In *Figure 1* the experimental excitation energy for the longest wave length transition of the anions and cations (= anion − 2 electrons) of acidic hydrocarbons (*Table X*) is plotted against the theoretical excitation energy calculated according to simple Hückel theory (*Table I*). Here it should be noted that most of the anion spectra were taken in solvents of high dielectric constant where the ions may be assumed to be essentially free.

As is known, simple Hückel theory is in general less adequate for predicting spectral data of ions of non-alternant hydrocarbons. Nevertheless, as can be seen in *Figure 1*, the correlation is surprisingly good. Similar observations have been made previously[108-113]. Particularly encouraging is the good agreement between theory and experiment for certain anion–cation pairs of non-alternant hydrocarbons[82]. It should be noted, however, that the good correlation breaks down (*cf. Figure 1*) if neutral molecules are also included (allowing for bond length alternation in the polyenes)[35]. Equally unsatisfactory is an attempt to include also cyanocarbon acids (*Figure 1*).

The conclusion to be drawn from these facts is that the Hückel theory may be used for a rough prediction of the longest wave length

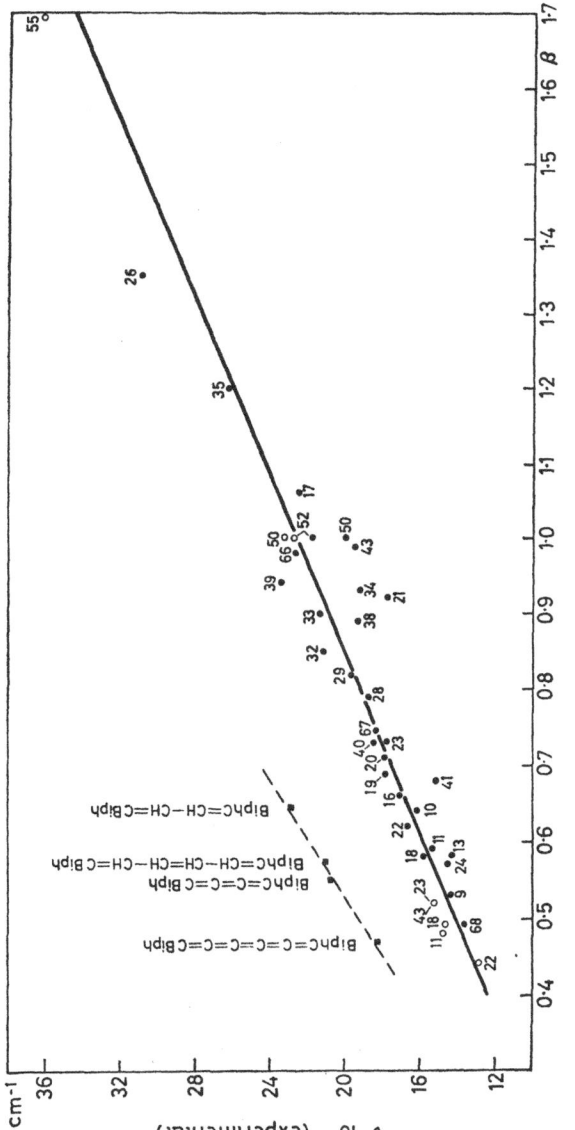

Figure 1. Correlation of experimental and theoretical excitation energies for anions (●) and cations (○) of acidic hydrocarbons. The numbers are those of Tables I and X. Experimental excitation energies: 66[107], 67[82], 68[82].

bands of the anions of unknown hydrocarbons. This may be of importance in preparative work where one wants to get a clue as to the possible colour of an anion to be synthesized. Hückel theory is certainly inadequate for a quantitative correlation.

The electron spin resonance-spectra of the radicals derived from the anions by loss of one electron have been analysed in some cases[82, 114]. The ultraviolet spectra of the radicals exhibit in general a much less intense absorption band at considerably longer wave lengths than the corresponding anions[82, 114]. In addition, at shorter wave lengths an absorption band of intensity comparable to the longest wave length band of the anions is observed. According to simple Hückel theory the longest wave length transition of alternant radicals is degenerate and the same as in the corresponding anions and cations. Inclusion of some electron repulsion in the theory will remove the degeneracy in the radicals, giving rise to a less intense absorption at longer wave lengths and a more intense absorption at shorter wave lengths[115]. Similar arguments apply also to non-alternant radicals. So the qualitative features of the ultra-violet spectra of the radicals are accounted for by theory.

Comparison of p_wK-values of Acidic Hydrocarbons with Calculations

In *Figure 2* the p_wK-values of *Table X* are plotted against the ΔM-values of *Table I*.

Although there is considerable scatter around the straight line (also noticed by STREITWIESER JR.[32]) the conclusions drawn on p. 128 are found to be qualitatively true. Particularly noteworthy is that the acid strength does indeed pass through a minimum in the series cyclopentadiene, indene, fluorene, 3,4 (or 1,2)-benzofluorene, 3,4,5,6 (or 1,2,7,8)-dibenzofluorene. However, the quantitative agreement between theory and experiment is clearly unsatisfactory and therefore an inquiry into the reasons for the major deviations from the straight line in *Figure 2* is now made (statistical corrections are neglected in view of the rather moderate correlation).

Steric Reasons—It is striking that in many of the compounds which lie above the straight line steric interference is present (24, 27, 31, 32, 34, 50). The simple Hückel calculations of *Table I* always refer to completely planar molecules. If some non-planarity in the anions in question is taken into account, smaller ΔM-values will be calculated, i.e. the corresponding points in the diagram will move to the left and thus improve the correlation.

144

Tautomeric Hydrocarbons—Several of the hydrocarbons listed in *Table X* may exist in tautomeric forms. For some of them only the gross acid strength is known and this is approximately the acid strength of the

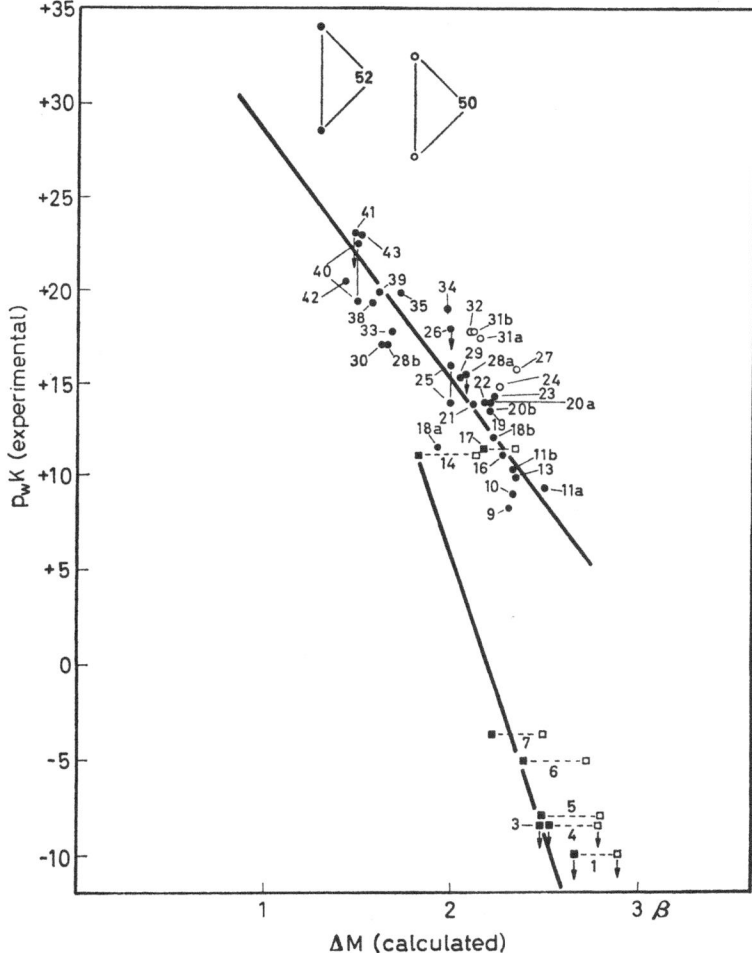

Figure 2. Correlation of experimental p_wK-values with ΔM-values calculated according to simple Hückel theory. ● *hydrocarbons,* ○ *sterically hindered hydrocarbons,* ■ *cyanocarbon acids* $(h_N=1)$, □ *cyanocarbon acids* $(h_N=2)$.

tautomer predominating at equilibrium. In five cases it has been possible to assign individual p_wK-values to each tautomer[51, 55]. Qualitative agreement between theory and experiment is evident from *Figure 2* for these tautomers. The most notable exception is bis-(fluorenylidene-methyl)methane (18a) which is predicted to be much less acidic than its tautomer 18b while in fact it is more acidic. This strong deviation is most likely due to an inadequate evaluation of the individual energies of the two tautomeric hydrocarbons. The anion being the same for both tautomers can be left out of consideration. Now the change in σ-bond energy and the change in solvation energy on dissociation were assumed to be constant (p. 130) so one has to inquire whether these two energy quantities are the same for both tautomers respectively. Adopting the σ-bond increments given by Dewar *et al.*[29] one calculates the σ-bond energies given in *Table XI*. There is only a small difference between the two tautomers. Moreover, 18a has actually more σ-bond energy than 18b which is opposite to that required. The same is true for the solvation energy (*Table XI*) calculated according to Jano's formula (which is derived from eqn 19) and using the Hückel net charges for q_t.

Table XI. σ-bond Energies and Solvation Energies for the Tautomeric Pentadienes 18a and 18b (Table X)

No.	Compound	$E_{\sigma b}^{(a)}$	$E_{solv}^{(a)}$
18a	BiphC=CH—CH$_2$—CH=CBiph	−5200·4	−2·6
18b	BiphC=CH—CH=CH—CHBiph	−5199·2	−2·0

(a) Energies in kcal/mole.

Finally, it is conceivable that the π-energy of the tautomers has been evaluated inadequately. The neglect of electron repulsion is certainly a major deficiency. However, remaining within the framework of simple Hückel theory, one might try whether inclusion of some bond length alternation will improve the situation. For the calculations reported in *Table I* all C—C bonds were taken to have the same length as the C—C bonds in benzene. This is certainly an oversimplification, especially for the hydrocarbons. One should adopt smaller resonance integrals for conjugated single bonds than for double bonds, i.e.

$$\beta_{rs} \text{ (single bond)} = k_{rs}\beta \qquad k_{rs} < 1$$

If p_{rs} and β_{rs} represent bond order and resonance integral of the bond r—s one has according to COULSON and LONGUET-HIGGINS[116]:

$$\Delta M = M(\text{R}^-) - 2\sum_{r<s} p_{rs}k_{rs}$$

146

Now in the tautomer 18a there are six conjugated single bonds the resonance integral of which should be reduced while in the tautomer 18b there are only five such single bonds. Thus inclusion of bond length alternation will increase the ΔM-value of tautomer 18a more than that of tautomer 18b and thereby will improve the correlation in *Figure 2*. There is only one other pair of tautomers listed in *Table X* for which the number of conjugated single bonds differs in the two tautomers, namely 28a and 28b. From *Figure 2* it is apparent that it is just these two pairs of tautomers (18 and 28) which show great deviations.

More advanced π-electron calculations automatically allow for bond length alternation, so it is regrettable that only very few[30] such calculations have so far been reported relating to acid/base equilibria.

Cyanocarbon Acids—Much too large p_wK-values are predicted for the cyanocarbon acids if the parameters of Boyd[37] are adopted for the calculation of ΔM:

$$h_N = 1, \qquad k_{CN} = 1{\cdot}2 \qquad (\alpha_N = \alpha_C + h_N\beta, \quad \beta_{CN} = k_{CN}\beta)$$

h_N is probably too low for the sp-hybridized nitrogen atoms in nitriles. For $h_N = 2$ the correlation is improved although these cyanocarbon acids are still much stronger acids than would have been anticipated from Hückel theory. It is possible that these deviations are due to solvation: the cyanocarbon acids are on an average much smaller than the hydrocarbons. From Hoijtink's formula (eqn (19)) it follows that the anions of cyanocarbon acids will be much better solvated than hydrocarbon anions.

An exception is 9-cyanofluorene the ΔM-value of which is not very sensitive to the choice of parameters. In addition, it fits the correlation quite well. This indicates that 9-cyanofluorene may be a good basis for the determination of p_wK-values of hydrocarbons.

In conclusion one may say that the prediction of acid strengths of hydrocarbons according to simple Hückel theory is only possible with moderate accuracy. However, one may anticipate that an SCF-calculation of the π-energy would considerably improve the agreement between theory and experiment, since bond length alternations and, to some extent, steric effects are automatically allowed for. A further improvement might be achieved by incorporating also the σ-bond energies, either from bond increments or, better still, from an SCF theory which takes into account all valence electrons. A more exact estimation of solvation energies seems still to be rather difficult.

147

SYNTHESES AND REACTIONS OF ACIDIC HYDRO-CARBONS AND CYANOCARBON ACIDS

The emphasis of this section lies on the presentation of general aspects. Syntheses and reactions of well known compounds are omitted.

Syntheses

The general idea in building up CH-acids consists in accumulating certain acidifying groups (e.g. fulvene or cyano groups) in one molecule so as to give maximum stability to the corresponding anion.

Syntheses via Methylene Compounds—Metal compounds of acidic hydrocarbons I add to halomethylene compounds II (available from a Wittig reaction with halomethylenetriphenylphosphorane) of acidic hydrocarbons to form new, in general, more acidic hydrocarbons III:

$$R_2CH^-M^+ + R'_2C{=}CHX \xrightarrow[\text{benzene}]{\text{DMSO or}} R'_2C{=}CH{-}CHR_2$$
$$\text{I} \qquad \text{II}: X = Cl, Br, I \qquad \text{III}$$

The condensation can be carried out in an aprotic solvent with excess hydroxide or alkoxide as base if hydrocarbons I are not much more acidic than fluorene. Otherwise the reaction is conducted in benzene with phenyllithium as base. In this case the free anion of III is not formed and the double bond is always found in the position indicated. By contrast the condensation in an aprotic solvent always involves the free anion of III. As a consequence mixtures of tautomers are frequently obtained upon acidification $(R {\neq} R')$. Compounds 11, 13, 16, 18, 20, 22, 27 (*Table X*) and further hydrocarbons (e.g. 69, 70) have been prepared in this way by KUHN et al.[87].

$$\begin{array}{cc} \text{BiphC}{=}\text{CH}{-}\text{CH(CH}{=}\text{CPh}_2)_2 & \text{(BiphC}{=}\text{CH)}_2\text{CH}{-}\text{CH}{=}\text{CPh}_2 \\ 69 & 70 \end{array}$$

The condensation of metal compounds of acidic hydrocarbons I with N,N-disubstituted aminomethylene compounds IV (available from the reaction of I with amidinium salts)[90, 117] proceeds along similar lines:

$$R_2CH^-M^+ + R'_2C{=}CH{-}NR''_2 \xrightarrow{\text{THF}} \text{III}$$
$$\text{I} \qquad \text{IV}$$

For $R_2CH_2 = R'_2CH_2 =$ indene or cyclopentadiene only the anions have so far been observed. Attempts to isolate the corresponding hydrocarbons led to polymerization. Compounds 11, 18 and 23 (*Table X*) and 71 were prepared from dimethylaminomethylene compounds. The latter is claimed to be the most acidic hydrocarbon so far known[118], but no acidity measurements were reported.

BiphC=CH—⬠—CH=CBiph $\xrightarrow{\text{BiphCHLi}}$ BiphC=CH—⬠—CH=CBiph

CH—N(CH$_3$)$_2$ ⟶ 71 CH—CHBiph

Syntheses via Pentyn-1,5-diols[119]—Ketones V of suitable acidic hydrocarbons I are transformed to pentyndiols VI by means of propargylbromide. After hydrogenation and double elimination of water pentadienes VII will result which are in general much more acidic than the starting hydrocarbons I:

$$R_2CH_2 \longrightarrow R_2C{=}O \xrightarrow[\text{Mg or Al}]{\text{HC}{\equiv}\text{C—CH}_2\text{Br}} \underset{\overset{|}{\text{OH}}}{R_2C}{-}CH_2{-}C{\equiv}CH \xrightarrow[\text{R}'_2\text{CO}]{\text{PhLi}}$$

$$\underset{\overset{|}{\text{OH}}}{R_2C}{-}CH_2{-}C{\equiv}C{-}\underset{\overset{|}{\text{OH}}}{CR'_2} \xrightarrow[(2)\ -2\text{H}_2\text{O}]{(1)\quad 2\text{H}_2} R_2C{=}CH{-}CH_2{-}CH{=}CR'_2$$

VI   VII

Instead of the 1,4-pentadienes VII isomeric 1,3-pentadienes have in some cases been obtained[87]. Compounds 9, 10, 15, 18 (*Table X*) were synthesized according to this method. Further ketones V utilized include benzophenone and tetraphenylcyclopentadienone. Halogenation of III and VII, respectively, and subsequent HX elimination leads to interesting cumulenes[97, 120].

Wawzonek Condensation[121]—The alcohols VIII and IX are likely to give the carbonium ions X and XI under acidic reaction conditions. Electrophilic substitution by X in the CH$_2$-group of XI would then produce the intermediate XII which may lose a proton to form acidic hydrocarbons XIII:

$$R_2CH{-}OH + \underset{\overset{|}{\text{OH}}}{\overset{\overset{\text{R}''}{|}}{CH_2}}{-}CR'_2 \xrightarrow[\text{AcOH}]{\text{H}^+} [R_2CH^+ + \overset{\overset{\text{R}''}{|}}{CH_2}{-}\overset{+}{C}R'_2] \xrightarrow{-\text{H}^+}$$

VIII   IX   X   XI

$$[R_2CH{-}\overset{\overset{\text{R}''}{|}}{CH}{-}\overset{+}{C}R'_2] \xrightarrow{-\text{H}^+} R_2CH{-}\overset{\overset{\text{R}''}{|}}{C}{=}CR'_2$$

XII   XIII

The double bond in XIII is always found in the position indicated. Compounds 11, 23, 24, 31b (*Table X*) and 69 and 70 may be obtained by this method[87].

Special Methods—Salts of cyclononatetraene (26) were obtained from the reaction of cyclooctatetraene with halomethanes[94]:

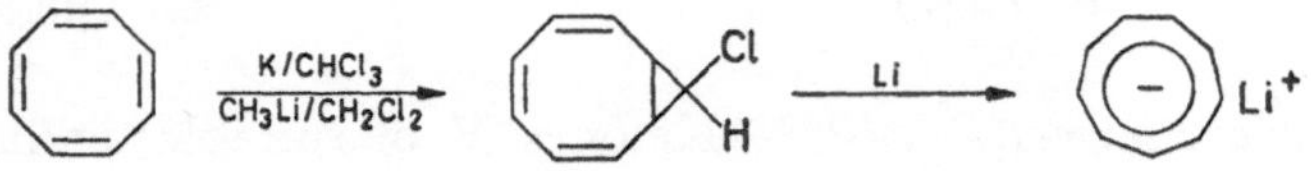

The free hydrocarbon has apparently not yet been isolated. Recently the corresponding bridged 1,5-methanocyclononatetraene was synthesized[150].

Fluoradene (21) is the main reaction product when 9-(o-aminophenyl)fluorene (72) is treated with nitrous acid[65]:

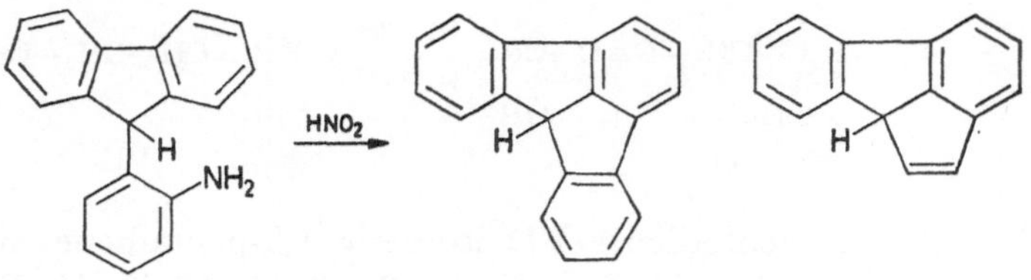

21 can also be obtained from 9-diazofluorene and benzyne in 65 per cent overall yield[151]. Attempts to prepare the desbenzo-derivate (73) of fluoradene have been unsuccessful[96, 122].

Syntheses of Cyanocarbon Acids—Most of these acids can only be isolated as salts (also hydronium salts) since the free acids do not even exist in concentrated sulphuric acid. Tricyanomethane (6) was obtained as free acid from bromomalononitrile and potassium cyanide[85]:

$$(NC)_2CHBr \xrightarrow{KCN} (NC)_3CH \text{ or } (NC)_2C{=}C{=}NH$$
$$6$$

$\alpha,\alpha,\omega,\omega$-Tetracyano-polyenes (74) result from the condensation of malononitrile with vinylogous formamides[123]:

$$\!\!>\!\!N{-}(CH{=}CH)_n{-}CHO \xrightarrow[(KOH)]{H_2C(CN)_2} [(NC)_2C{\cdots}CH{\cdots}(CH{\cdots}CH)_n{\cdots}CCN)_2]^-K^+$$
$$74, n = 0, 1, 2$$

Pentacyanopropene (4) is a condensation product of tetracyanoethylene under basic conditions[84]:

$$2(NC)_2C{=}C(CN)_2 + 2H_2O \longrightarrow (NC)_2C{=}C{-}CH(CN)_2 + CO_2 + 3HCN$$
$$4 \quad\quad\; CN$$

Hexacyanoisobutene (3) a dibasic acid is obtained from dicyano-ketene acetal and malononitrile[84]:

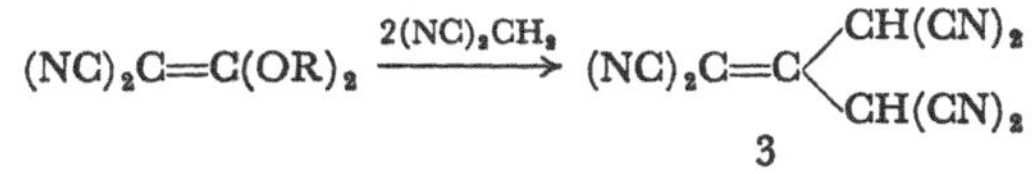

$$(NC)_2C{=}C(OR)_2 \xrightarrow{2(NC)_2CH_2} (NC)_2C{=}C\Big\langle \begin{smallmatrix} CH(CN)_2 \\ CH(CN)_2 \end{smallmatrix}$$

3

2-Cyanomethyl-1,1,3,3-tetracyanopropene is a self-condensation product of malononitrile[123a]. This compound like 3 and 4 can only be isolated as salts.

Hexacyanobutendiide (75) cyclizes in dilute aqueous solution. Hydrolysis and decarboxylation yields tetracyano-amino-cyclopentadienide (76) which may be converted into tetracyano- and pentacyano-cyclopentadienide as shown[83]:

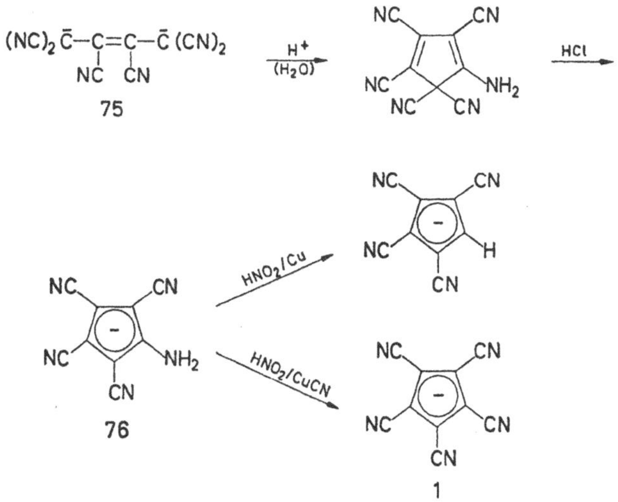

All the various cyano-substituted cyclopentadienides can be obtained from cyclopentadiene and cyanogen chloride[152]. The potassium salt of 1 can be heated at 400° in air without decomposition. The acid of 1 is reported to be 10^{10} times more acidic than sulphuric acid in acetonitrile.

Trisdicyanovinylcyclopentadienide[124] (78) results from the reaction of malononitrile with the dialdehyde (77).

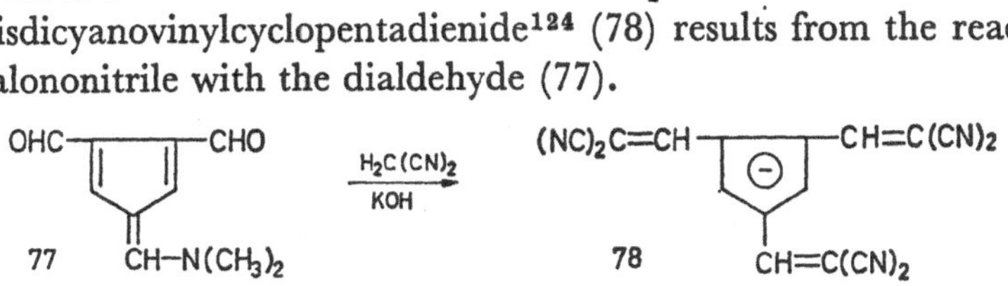

Reactions

The strong cyanocarbon acids recorded in *Table X* are chemically rather inert. Many of them are not even hydrolysed in 80% sulphuric acid.

In most cases the free acids are unknown and only various salts can be isolated. The salt (79) is of interest as it is composed of a carboniumion and a carbanion[80]:

$$[(NC)_2C \equiv C \equiv C(CN)_2]^- [(p\text{-}CH_3O\text{---}C_6H_4)_3C]^+$$
$$\overset{|}{C}N \qquad 79$$

2-substituted tetracyanopropenides have been found to cyclize to cyanosubstituted pyridines[125]. Derivatives of tetracyanocyclopentadienide were obtained by electrophilic substitution[153]. BOYD et al.[125a] have performed an extensive study of the thermochemistry of cyanocarbon acids and salts.

Somewhat more space will be devoted to the chemistry of hydrocarbon acids.

Formation of Carbanions—Hydrocarbons with $p_wK < 30$ may be deprotonized by base in aprotic solvents to give in general deeply coloured carbanions. Particularly useful are dipolar aprotic solvents such as dimethylformamide (DMF), dimethylsulphoxide (DMSO) and hexamethylphosphortriamide (HMPT) in which case it is sufficient to use hydroxide or alkoxide as base. Thus bis- and tris-(2,2-diphenylvinyl) methane are smoothly converted into carbanions by potassium-*t*-butoxide in DMSO[51] while in ether phenyllithium or phenylisopropyl-potassium[126] are required. The most acidic hydrocarbons recorded in *Table X* are completely deprotonized by tripropylamine in DMSO[87]. By contrast to the situation obtaining with solvents used earlier in carbanion-chemistry there are apparently essentially free carbanions in the dipolar aprotic solvents mentioned. This is well exemplified by a recent observation[127] that benzylmagnesium chloride prepared in ether develops a deep red colour if HMPT is added. This was ascribed to the presence of the free benzyl anion.

Protolytic reactions of hydrocarbons in non-aqueous solvents are frequently not instantaneous. The rate of anion formation depends on the solvent, on the base strength and on the structure of the hydrocarbon. The rate of metalation is greatly enhanced in polar solvents and increases steadily in the series heptane, benzene, tetrahydrofuran, dimethoxyethane, DMF, DMSO, HMPT[128]. The rate of anion formation does not parallel the acid strength of the hydrocarbons in some cases. Thus fluoradene (21) is deprotonated much faster than 9-(fluorenylidenemethyl)fluorene (23) in DMSO/tripropylamine although both hydrocarbons are of comparable acid strength.[51] This finding is likely to be attributable to the fact that the geometry of fluoradene is already much closer to the geometry of its anion than is the case with hydrocarbon 23. Rates of proton transfer reactions involving the anions of acidic hydrocarbons are currently under active investigation[154].

Reactions of Carbanions—Most carbanions are sensitive to oxygen. The stability towards oxygen strongly depends on solvent[129] and is greatly increased in dipolar aprotic solvents. The stability of the anions towards oxygen roughly parallels the acid strength of the corresponding hydrocarbons. The stability passes through a minimum at fluorene in a series of analogous propenes containing terminal cyclopentadiene-, indene-, fluorene- and dibenzofluorene-groups[87]. In a series of substituted propenes, pentadienes and heptatrienes the heptatrienes are the least stable towards oxygen[117]. A measure for the sensitivity towards oxygen is the ionisation potential of the anions, a theoretical estimate of which is provided by the highest occupied molecular orbital in the anion[130]. Comparing column 3 of *Table I* with the order of stability just mentioned a rough correlation is clearly discernible.

The oxidation of certain carbanions leads to stable radicals[82, 91, 114]. Dimers and peroxides thereof were also observed. In some cases disproportionation of the radicals gives rise to the corresponding cations[82], for example:

$$[BiphC\text{---}CH\text{---}CH\text{---}CH\text{---}CBiph]^{-}K^{+}$$

$$\downarrow \text{-}e^{-}$$

$$[BiphC\text{---}CH\text{---}CH\text{---}CH\text{---}CBiph]^{\bullet} \rightleftharpoons Dimer$$

$$\downarrow HClO_4/AcOH$$

$$[BiphC\text{---}CH\text{---}CH\text{---}CH\text{---}CBiph]^{+}ClO_4^{-} + \text{ other products}$$

Many of the hydrocarbons recorded in *Table X* form ambident carbanions. Subsequent protonation[51, 89, 97] or alkylation[51, 91, 131–133] may lead to isomers (*cf. Table XII*):

$$\overset{R}{\underset{|}{C}}-C=C \quad \xrightarrow{k_1} \quad \left[C\text{···}C\text{···}C \right]^{-} \quad \xrightarrow{k_2} \quad C=C-\overset{}{\underset{R}{C}}$$

As can be seen from *Table XII*, the ratio of the isomers formed corresponds qualitatively with the Hückel π-electron charge distribution[133a]. A comparison with *Table IX* reveals that the equilibrium ratio of tautomeric hydrocarbons differs strongly from the corresponding ratio obtained by kinetically controlled protonation of the anion. Thus acidification of 9-styrylfluorenide in DMSO with acetic acid yields 70%

of trans-9-styrylfluorene which subsequently is slowly transformed into 9-phenylethylidenefluorene[51]:

$$[BiphC \dddot{-} CH \dddot{-} CH{-}Ph]^{-}K^{+} \xrightarrow[\text{fast}]{\underset{DMSO}{H^{+}}} BiphCH{-}\overset{H}{\underset{H}{C}}{=}\overset{}{C}{-}Ph \xrightarrow{\text{slow}} BiphC{=}CH{-}CH_2{-}Ph$$

$$\qquad\qquad\qquad\qquad\qquad\qquad\qquad 28a \qquad\qquad\qquad\qquad 28b$$

Table XII. Formation of Isomers on Kinetically Controlled Protonation or Alkylation of Ambident Carbanions in DMF or DMSO[51]; π-electron Charge Densities from Simple Hückel Theory at the Positions in Question[35]

Carbanion[a]	Yield (%)			
	Protonation		Methylation	
	C_1	C_3	C_1	C_3
1·22 1·18 $[BiphC \text{---} CH \text{---} CH{-}Ph]^{-}$ 1 3	≥ 70	≤ 30	> 80	< 20
1·19 1·16 $[BiphC \text{---} CH \text{---} CPh_2]^{-}$ 1 3	93	7	96	4
1·15 1·14 $[3,4,5,6\text{-}BiphC \text{---} CH \text{---} CBiph]^{-}$ 1 3	> 90	< 10	> 90	< 10
1·14 1·14 $[1,2,7,8\text{-}BiphC \text{---} CH \text{---} CBiph]^{-}$ 1 3	100	0	100	0
1·17 1·17 $[BiphC \text{---} CH \text{---} CPhen]^{-}$ 1 3	60	40	55	45
1·15 1·13 $[BiphC \text{---} CH \text{---} CH \text{---} CH \text{---} CBiph]^{-}$ 1 3	76	24		

(a) Numbers above the formula indicate Hückel-π-charge densities.

This behaviour is reminiscent of similar phenomena encountered in the field of keto/enol-, nitro/acinitro- or azapropene-tautomerism[134]. The HUGHES–INGOLD rule[135] states that of two tautomers the thermodynamically less stable is formed most quickly on acidification of a solution of the common ion. In addition, the less stable isomer should always be kinetically the more acidic. As Eigen[18] has emphasized, this cannot be generally true since the rate of protonation is governed by the pK-difference between proton donor and acceptor and therefore the less acidic tautomer should be favoured on protonation of the anion. If the two tautomers are CH- and OH- or NH-acids, respectively, the

Hughes–Ingold rule is usually fulfilled, i.e. the less stable but kinetically more acidic OH- or NH-acids are formed in a kinetically controlled protonation. However, this is most likely due to the much better hydrogen bonding ability of the heteroatoms. For ambient anions which contain no heteroatoms the situation may be altogether different. Thus CRAM et al.[136-138] investigated several allylic systems (phenylsubstituted propenes and butenes) and concluded through a combination of kinetics of isomerization, the thermodynamic data and kinetics of H/D-exchange that one or both statements of the Hughes–Ingold rule may be invalid (depending on the particular activation energy/reaction coordinate profiles).

A somewhat less elaborate technique of measurement is applicable if stronger hydrocarbon acids are investigated, for example[55].

$$BiphC\underset{H}{-}CH=C\overset{R}{\underset{R'}{}} \xrightleftharpoons[k_{-1}]{k_1} \left[BiphC\cdots CH\cdots C\overset{R}{\underset{R'}{}}\right]^{-} \xrightleftharpoons[k_{-2}]{k_2} BiphC=CH\underset{H}{-}C\overset{R}{\underset{R'}{}}$$

20a: R—R′=Phen	20b: R—R′=Phen
28a: R=H, R′=Ph	28b: R=H, R′=Ph
31a: R=R′=Ph	31b: R=R′=Ph

From *Table IX* the following equilibrium ratios are found.

$$\frac{20a}{20b} = 2\cdot85 \qquad \frac{28a}{28b} \leq 0\cdot01 \qquad \frac{31a}{31b} = 0\cdot35$$

On the other hand, from the ratio of tautomers produced by a kinetically controlled protonation one can deduce (*Table XII*)

$$20: \quad \frac{k_{-1}}{k_{-2}} = 1\cdot5 \qquad 28: \quad \frac{k_{-1}}{k_{-2}} = 2\cdot32 \qquad 31: \quad \frac{k_{-1}}{k_{-2}} = 13\cdot3$$

$$\frac{k_1}{k_2} \simeq 0\cdot5 \qquad\qquad \frac{k_1}{k_2} \simeq 250 \qquad\qquad \frac{k_1}{k_2} \simeq 38$$

An independent confirmation of these results comes from the observation that 20b, 28a and 31a exhibit a higher rate of anion formation than their respective tautomers.[55] These findings are in accord with the Hughes–Ingold rule in the case of the tautomers 28 and 31. For the tautomers 20 the rule is only partially fulfilled: the thermodynamically less stable tautomer (20b) is kinetically more acidic, but the more stable tautomer 20a is predominantly produced by the kinetically controlled protonation of the anion.

Isomerization of Acidic Hydrocarbons—Base catalysed and thermal isomerization are well-known reactions in the field of acidic hydrocarbons. Many of the strong hydrocarbon acids synthesized by KUHN et al. undergo facile isomerization[51, 92] even in the presence of mineral acids, but nothing is known as yet about the detailed reaction mechanism. By contrast, recent investigations into the kinetics of isomerization and hydrogen isotope exchange of weaker hydrocarbon acids, together with a study of the stereochemical course of these reactions, led to important results as to the mechanism of isomerization and the geometric stability of carbanions. The material has been excellently reviewed recently by CRAM[139] and therefore only two main results of these investigations are briefly mentioned here: (i) the intramolecularity of base catalysed[136, 140-145] and thermal[132, 143, 146] isomerizations; and (ii) the stereoselectivity of isomerizations[137, 139, 147] and other reactions which involve carbanionic intermediates (e.g. various substitution reactions of acidic hydrocarbons and organometallic compounds). The conclusion drawn from these observations is that the carbanionic intermediates (complexed by solvent and base) have a definite structure and some geometric stability. For example, the carbanion intermediate XIV was postulated by CRAM and UYEDA[136] in order to explain the intramolecularity of the isomerization of optically active 3-phenyl-1-butene and the stereochemical course of the H/D-exchange of the starting material. The protonated form of the base is hydrogen bonded to both ends of the allylic system and is placed on one side of the plane made up by the carbanion.

XIV

REFERENCES

[1] BERTHELOT, G. Ann. Chim. [4] 9 (1866) 385; MATIGNON, C. C.R. Acad. Sci., Paris 124 (1897) 775
[2] GLASER, C. Liebigs Ann. 154 (1870) 137
[3] HANRIOT, M. and SAINT-PIERRE, O. Bull. Soc. Chim. Fr. [3] 1 (1889) 774
[4] SAINT-PIERRE, O. Bull. Soc. Chim. Fr. [3] 5 (1891) 292
[5] THIELE, J. Ber. dtsch. chem. Ges. 34 (1901) 68
[6] WEISSGERBER, R. Ber. dtsch. chem. Ges. 34 (1901) 1659
[7] JONES, H. C. and ALLEN, C. R. Amer. chem. J. 18 (1896) 1; WALKER, J. and CORMACK, W. J. chem. Soc. 77 (1900) 19; BILLITZER, J. Monatsh. Chem. 23 (1902) 489, 502
[8] WOOSTER, C. B. and MITCHELL, N. W. J. Amer. chem. Soc. 52 (1930) 688

[9] WOOSTER, C. B. *Chem. Rev.* 11 (1932) 1

[10] ZIEGLER, K. *Angew. Chem.* 49 (1936) 455

[11] CONANT, J. B. and WHELAND, G. W. *J. Amer. chem. Soc.* 54 (1932) 1212

[12] McEWEN, W. K. *J. Amer. chem. Soc.* 58 (1936) 1124

[13] HÜCKEL, E. *Z. Phys.* 70 (1931) 204

[14] WHELAND, G. W. *J. chem. Phys.* 2 (1934) 474

[15] BRÖNSTED, J. N. and PEDERSEN, K. J. *Z. phys. Chem.* 108 (1924) 185

[16] BELL, R. P. *The Proton in Chemistry*, p. 161, Cornell University Press, New York, 1959

[17] BELL, R. P. *The Proton in Chemistry*, p. 163, Cornell University Press, New York, 1959

[18] EIGEN, M. *Angew. Chem.* 75 (1963) 489

[19] ARNDT, F. *Abh. Braunschw. wiss. Ges.* 8 (1956) 1

[20] BRÖNSTED, J. N. *Recl. Trav. chim.* 42 (1923) 718

[21] PARKER, A. J. *Quart. Rev. chem. Soc.* 16 (1962) 163–187; compare also KOLTHOFF, I. M. and REDDY, T. B. *Inorg. chem.* 1 (1962) 189

[22] BELL, R. P. *The Proton in Chemistry*, p. 54, Cornell University Press, New York, 1959

[23] KOLTHOFF, I. M. and BRUCKENSTEIN, S. *J. Amer. chem. Soc.* 78 (1956) 1

[24] CRAM, D. J. *Fundamentals of Carbanion Chemistry*, p. 49, Academic Press, New York and London, 1965

[25] DESSY, R. E., OKUZUMI, Y. and CHEN, A. *J. Amer. chem. Soc.* 84 (1962) 2899

[26] EVANS, M. G. and POLANYI, M. *Trans. Faraday Soc.* 32 (1936) 1333

[27] BELL, R. P. *The Proton in Chemistry*, p. 71, Cornell University Press, New York, 1959

[28] LEFFLER, J. E. *J. org. Chem.* 20 (1955) 1202

[29] DEWAR, M. J. S. and SCHMEISING, H. N. *Tetrahedron* 11 (1960) 96;—and CHUNG, A. L. H. *J. chem. Phys.* 42 (1965) 756

[29a] POPLE, J. A. and SEGAL, J. A. *J. chem. Phys.* 44 (1966) 3289; DEWAR, M. J. S. and KLOPMAN, G. *J. Amer. chem. Soc.* 89 (1967) 3089

[30] CHALVET, O., DAUDEL, R. and PERADEJORDI, F. in *Molecular Orbitals in Chemistry, Physics and Biology* (LÖWDIN, P. O. and PULLMAN, B.), p. 475, Academic Press, New York and London, 1964

[31] DEWAR, M. J. S. and GLEICHER, G. J. *J. Amer. chem. Soc.* 87 (1965) 685, 692

[32] STREITWIESER, A. *Tetrahedron Lett.* 6 (1960) 23

[33] STREITWIESER, A. *Molecular Orbital Theory for Organic Chemists*, p. 415, Wiley, New York and London, 1961

[34] STREITWIESER, A. and BRAUMAN, J. I. *Supplemental Tables of Molecular Orbital Calculations*, 1965, Pergamon Press, Oxford

[35] FISCHER, H. and EGE, G. unpublished calculations

[36] ZAHRADNIK, R., MICHL, J. and KOUTECKY, J. *Colln Czech. chem. Commun.* 29 (1964) 1932

[37] BOYD, R. H. *J. phys. Chem.* 65 (1961) 1834

[38] KING, E. J. *Acid Base Equilibria*, Vol. 4 in *The International Encyclopedia of Physical Chemistry and Chemical Physics* (ed. ROBINSON, R. A.), p. 280, Pergamon Press, Oxford, 1965; GLOVER, D. J. *J. Amer. chem. Soc.* 87 (1965) 5275, 5279

[39] BORN, M. *Z. Phys.* 1 (1920) 45

[40] HOIJTINK, G. J., DE BOER, E., VAN DER MEIJ, P. M. and WEIJLAND, W. P. *Recl. Trav. chim.* 75 (1956) 487

[41] Jano, I. *C.R. Acad. Sci., Paris* 261 (1965) 103

[42] Streitwieser, A., Brauman, J. I., Hammons, J. H. and Pudjaatmaka, A. H. *J. Amer. chem. Soc.* 87 (1965) 384; and Ciuffarin, E., *ibid.* 89 (1967) 59; *ibid.* 89 (1967) 63

[43] Steiner, E. C. and Gilbert, J. M. *J. Amer. chem. Soc.* 87 (1965) 382; see, however, Steiner, E. C. and Starkey, J. D. *J. Amer. chem. Soc.* 89 (1967) 2751 for a correction of the previous results

[44] Ziegler, K. and Wollschitt, H. *Liebigs Ann.* 479 (1930) 123

[45] Streitwieser, A., Padgett, W. M. and Schwager, I. *J. phys. Chem.* 68 (1964) 2922

[46] Dixon, J. A., Gwinner, P. A. and Lini, D. C. *J. Amer. chem. Soc.* 87 (1965) 1379

[47] Hogen-Esch, T. E. and Smid, J. *J. Amer. chem. Soc.* 87 (1965) 669; *ibid.* 88 (1966) 307

[48] Szwarc, M. *Makromolek. Chem.* 89 (1965) 44

[49] Buschow, K. H. J., Dieleman, J. and Hoijtink, G. J. *J. chem. Phys.* 42 (1965) 1993; Velthorst, N. H. and Hoijtink, G. J. *J. Amer. chem. Soc.* 87 (1965) 4529

[50] Slates, R. V. and Szwarc, M. *J. phys. Chem.* 69 (1967) 4124

[51] Kuhn, R. and Rewicki, D. *Liebigs Ann.* 704 (1967) 9; *ibid.* 706 (1967) 250 and unpublished work

[52] Bowden, K. and Stewart, R. *Tetrahedron* 21 (1965) 261

[53] Langford, C. H. and Burwell, R. L. *J. Amer. chem. Soc.* 82 (1960) 1503

[54] Deno, N. C. *J. Am. chem. Soc.* 74 (1952) 2039

[55] Kuhn, R. and Rewicki, D. *Tetrahedron Lett.* (1965) 3513

[56] Stewart, R., O'Donnell, J. P., Cram, D. J. and Rickborn, B. *Tetrahedron* 18 (1962) 917

[57] Steiner, E. C. and Gilbert, J. M. *J. Amer. chem. Soc.* 85 (1963) 3054

[58] Schriesheim, A. and Rowe, C. A. *J. Amer. chem. Soc.* 84 (1962) 3161

[59] Cram, D. J., Rickborn, B. and Knox, G. R. *J. Amer. chem. Soc.* 82 (1960) 6412; Cram, D. J., Rickborn, B., Kingsbury, C. A. and Haberfield, P. *J. Amer. chem. Soc.* 83 (1961) 3678

[60] Cram, D. J., Kingsbury, C. A. and Rickborn, B. *J. Amer. chem. Soc.* 83 (1961) 3688

[61] Stearns, R. S. and Wheland, G. W. *J. Amer. chem. Soc.* 69 (1947) 2025

[62] Boyd, R. H. *J. phys. Chem.* 67 (1962) 737

[63] Applequist, D. E. and O'Brien, D. F. *J. Amer. chem. Soc.* 85 (1963) 743

[64] Salinger, R. M. and Dessy, R. E. *Tetrahedron Lett.* (1963) 729; Dessy, R. E., Kitching, W., Psarras, T., Salinger, R., Chen, A. and Chivers, T. *J. Amer. chem. Soc.* 88 (1966) 460

[65] Rapoport, H. and Smolinsky, G. *J. Amer. chem. Soc.* 82 (1960) 934

[66] Shatenshtein, A. I. *Adv. phys. org. Chem.* (Gold, V.), Academic Press, London and New York, Vol. 1 (1963) 156; compare also Streitwieser, A., Caldwell, R. A., Lawler, R. G. and Ziegler, G. R. *J. Amer. chem. Soc.* 87 (1965) 5399 and earlier papers

[67] Lewis, G. N. and Seaborg, G. T. *J. Amer. chem. Soc.* 75 (1953) 1894; Caldin, E. F. and Trickett, J. C. *Trans. Faraday Soc.* 49 (1953) 772

[68] Pearson, R. G. and Dillon, R. L. *J. Amer. chem. Soc.* 75 (1953) 2439

[69] Shatenshtein, A. I. *Dokl. Akad. Nauk SSSR* 60 (1949) 1029

[69a] Ritchie, C. D. and Uschold, R. E. *J. Amer. chem. Soc.* 89 (1967) 1721; *ibid.* 89 (1967) 2752

[70] STREITWIESER, A., LANGWORTHY, W. C. and BRAUMAN, J. I. *J. Amer. chem. Soc.* 85 (1963) 1761

[71] HAMMETT, L. P. *Physical Organic Chemistry*, McGraw-Hill, New York (1940)

[72] HAMMETT, L. P. and DEYRUP, A. J. *J. Amer. chem. Soc.* 54 (1932) 2721

[73] PAUL, M. A. and LONG, F. A. *Chem. Rev.* 57 (1957) 1

[74] LEFFLER J. E. and GRUNDWALD, E. *Rates and Equilibria of Organic Reactions*, Chapter 8, John Wiley and Sons, New York (1963)

[75] STEWART, R. and O'DONNELL, J. P. *Can. J. Chem.* 42 (1964) 1681

[76] CRAM, D. J. *Chem. and Engng. News* (1963) 92

[77] STREITWIESER, A. and LAWLER, R. G. *J. Amer. chem. Soc.* 87 (1965) 5388

[78] CRAM, D. J. *Fundamentals of Carbanion Chemistry*, p. 19, Academic Press, New York and London, 1965

[79] BALLINGER, P. and LONG, F. A. *J. Amer. chem. Soc.* 82 (1960) 795; ARNETT, E. M. in *Progress in Physical Organic Chemistry* (ed. COHEN, S. G., STREITWIESER, A. and TAFT, R. W.), Vol. 1, p. 353, John Wiley and Sons, London and New York, 1963

[80] BOYD, R. H. *J. Amer. chem. Soc.* 85 (1963) 1555

[81] JONES, H. T. and WEISSMAN, S. I. *J. Amer. chem. Soc.* 84 (1962) 4269

[82] KUHN, R. and REWICKI, D. *Liebigs Ann.* 690 (1966) 50

[83] WEBSTER, O. W. *J. Amer. chem. Soc.* 87 (1965) 1820

[84] MIDDLETON, W. J., LITTLE, E. L., COFFMAN, D. D. and ENGELHARDT, V. A. *J. Amer. chem. Soc.* 80 (1958) 2795

[85] COX, E. and FONTAINE, A. *Bull. Soc. chim. Fr.* (1954) 948; TROFIMENKO, S. *J. org. Chem.* 28 (1963) 217

[86] WILLIAMS, J. K. *J. Amer. chem. Soc.* 84 (1962) 3478

[87] KUHN, R., FISCHER, H., REWICKI, D. and FISCHER, H. *Liebigs Ann.* 689 (1965) 1

[88] KUHN, R. and FISCHER, H. *Angew. Chem.* 76 (1964) 146

[89] KUHN, R. and REWICKI, D. *Tetrahedron Lett.* (1964) 383

[90] JUTZ, C. and AMSCHLER, H. *Angew. Chem.* 76 (1964) 302

[91] KUHN, R., FISCHER, H., NEUGEBAUER, F. A. and FISCHER, H. *Liebigs Ann.* 654 (1962) 64

[92] KUHN, R. *Angew. Chem.* 75 (1963) 870

[93] KUHN, R. *Angew. Chem.* 75 (1963) 174

[94] KATZ, T. J. and GARRATT, P. J. *J. Amer. chem. Soc.* 85 (1963) 2852; *ibid.* 86 (1964) 5194; LALANCETTE, E. A. and BENSON, R. E. *J. Amer. chem. Soc.* 85 (1963) 2853; *ibid.* 87 (1965) 1941; SIMMONS, H. E., CHESNUT, D. B. and LALANCETTE, E. A. *J. Amer. chem. Soc.* 87 (1965) 982

[95] MARTIN, R. H. *J. chem. Soc.* (1941) 679

[96] KUHN, R. and BREYER, U. *Liebigs Ann.* 661 (1963) 173; BREYER, U. *Ph.D. Thesis*, Heidelberg, 1962

[97] FISCHER, H. and FISCHER, H. *Chem. Ber.* 97 (1964) 2975

[98] SCHOEPFLE, C. S. *J. Amer. chem. Soc.* 44 (1922) 188

[99] TSCHITSCHIBABIN, A. E. and MAGIDSON, O. J. *J. prakt. Chem.* [2] 90 (1914) 168

[100] ASTAF'EV, J. V. and SHATENSHTEIN, A. I. *Optics and Spectroscopy* (English transl.) 6 (1958) 410

[101] WOODING, N. S. and HIGGINSON, W. C. *J. chem. Soc.* (1952) 774

[102] KOLTHOFF, I. M., BRUCKENSTEIN, S. and CHANTOONI, M. K. *J. Amer. chem. Soc.* 83 (1961) 3927

[103] BELL, R. P. *The Proton in Chemistry*, p. 93, Cornell University Press, New York, 1959

[104] RIEMENSCHNEIDER, R. and SCHÖNFELDER, R. *Z. Naturf.* 18b (1963) 979

[105] BIRCHALL, T. and JOLLY, W. L. *J. Amer. chem. Soc.* 87 (1965) 3007

[106] GAUTIER, J. A., MIOCQUE, M. and MOSKOWITZ, H. *C.R. Acad. Sci., Paris* 260 (1965) 1988

[107] REWICKI, D. *Chem. Ber.* 99 (1966) 392

[108] MEUCHE, D., STRAUSS, H. and HEILBRONNER, E. *Helv. chim. Acta* 41 (1958) 57, 414; NAVILLE, G., STRAUSS, H. and HEILBRONNER, E., *ibid.* 43 (1960) 1221

[109] STREITWIESER, A. and BRAUMAN, J. I. *J. Amer. chem. Soc.* 85 (1963) 2633

[110] FISCHER, H. *Thesis submitted to the University of Heidelberg*, in partial fulfilment of the requirements for a lectureship (1964)

[111] KOUTECKY, J., HOCHMAN, P. and MICHL, J. *J. chem. Phys.* 40 (1964) 2439

[112] ZAHRADNIK, R. *Angew. Chem.* 77 (1965) 1097

[113] BOYD, G. V. and SINGER, N. *Tetrahedron* 22 (1966) 547

[114] KUHN, R. and NEUGEBAUER, F. A. *Monatsh Chem.* 95 (1964) 3

[115] LONGUET-HIGGINS, H. C. and POPLE, J. A. *Proc. phys. Soc.* A 68 (1955) 591

[116] COULSON, C. A. and LONGUET-HIGGINS, H. C. *Proc. Roy. Soc.* 191 (1947) 39

[117] JUTZ, C. and AMSCHLER, H. *Angew. Chem.* 73 (1961) 806

[118] JUTZ, C. *Angew. Chem.* 77 (1965) 224

[119] MARIN, G., CHODKIEWICZ, W., CADIOT, P. and WILLEMART, A. *Bull. Soc. chim. Fr.* (1958) 1594

[120] KUHN, R., FISCHER, H. and FISCHER, H. *Chem. Ber.* 97 (1964) 1760

[121] WAWZONEK, S. and DUFEK, E. *J. Amer. chem. Soc.* 78 (1956) 3530

[122] McDOWELL, B. L., SMOLINSKY, G. and RAPOPORT, H. *J. Amer. chem. Soc.* 84 (1962) 3531

[123] STRELL, M., BRAUNBRUCK, W. B., FÜHLER, W. F. and HUBER, O. *Liebigs Ann.* 587 (1954) 177

[123a] KELLY, R. B., SLOMP, G. and CARON, E. L. *J. org. Chem.* 30 (1965) 1036

[124] MATSUMURA, S. and SETO, S. *Chem. pharm. Bull., Tokyo* 11 (1963) 126; *Chem. Abstr.* 59 (1963) 5071g

[125] MIDDLETON, W. J., ENGELHARDT, V. A. and FISHER, B. S. *J. Amer. chem. Soc.* 80 (1958) 2822

[125a] BOYD, R. H. *J. chem. Phys.* 38 (1963) 2529; BOYD, R. H. and WANG, C.-H. *J. Amer. chem. Soc.* 87 (1965) 430; RANJANGUHA, K. and WUTHRICH, R. *J. phys. Chem.* 71 (1967) 2178

[126] WITTIG, G. and OBERMANN, B. *Ber. dtsch. chem. Ges.* 68 (1935) 2214;— and KOSACK, H. *Liebigs Ann.* 529 (1937) 167

[127] EBEL, H. F. and SCHNEIDER, R. *Angew. Chem.* 77 (1965) 914

[128] OKHLOBYSTIN, O. YU. and ZAKHARKIN, L. I. *J. organometall. Chem.* 3 (1965) 257; NORMANT, H. *et al. Bull. Soc. chim. Fr.* (1965) 1872, 1881, 3441, 3446; FRAENKEL, G., ELLIS, S. H. and DIX, D. T *J. Amer. chem. Soc.* 87 (1965) 1406

[129] PACIFICI, J. G., GARST, J. F. and JANZEN, E. J. *J. Amer. chem. Soc.* 87 (1965) 3014

[130] STREITWIESER, A. *Molecular Orbital Theory for Organic Chemists*, p. 185, Wiley, New York and London, 1961

[131] McLEAN, S. and HAYNES, P. *Tetrahedron* 21 (1965) 2313

[132] MIRONOV, V. A., SOBOLEV, E. V. and ELIZAROVA, A. N. *Tetrahedron* 19 (1963) 1939

[133] FISCHER, H. and FISCHER, H. *Chem. Ber.* 99 (1966) 658

133a STREITWIESER, A. *Molecular Orbital Theory for Organic Chemists*, p. 425, Wiley, New York and London, 1961; STREITWIESER, A. and SUZUKI, S. *Tetrahedron* 16 (1961) 153

134 INGOLD, C. K. *Structure and Mechanism in Organic Chemistry*, Chap. 10, Cornell University Press, New York, 1953

135 INGOLD, C. K. *Structure and Mechanism in Organic Chemistry*, p. 565, Cornell University Press, New York, 1953

136 CRAM, D. J. and UYEDA, R. T. *J. Amer. chem. Soc.* 84 (1962) 4358; *ibid.* 86 (1964) 5466

137 HUNTER, D. H. and CRAM, D. J. *J. Amer. chem. Soc.* 86 (1964) 5478

138 CRAM, D. J. *Fundamentals of Carbanion Chemistry*, p. 209, Academic Press, New York and London, 1965

139 CRAM, D. J. *Fundamentals of Carbanion Chemistry*, Chaps. 3–5, Academic Press, New York and London, 1965

140 BANK, S., ROWE JR., C. A. and SCHRIESHEIM, C. A. *J. Amer. chem. Soc.* 85 (1963) 2115

141 DOERING, W. v. E. and GASPAR, P. P. *J. Amer. chem. Soc.* 85 (1963) 3043

142 BERGSON, G. and WEIDLER, A. M. *Acta chem. scand.* 17 (1963) 862, 1798, 2691, 2724

143 McLEAN, S. and HAYNES, P. *Tetrahedron Lett.* (1964) 2385;—*Tetrahedron* 21 (1965) 2329

144 CRAM, D. J., WILLEY, F., FISCHER, H. P. and SCOTT, D. A. *J. Amer. chem. Soc.* 86 (1964) 5370

145 BATES, R. B., CARNIGHAN, R. H. and STAPLES, C. E. *J. Amer. chem. Soc.* 85 (1963) 3032

146 ROTH, W. R. *Tetrahedron Lett.* (1964) 1009

147 BANK, S., SCHRIESHEIM, A. and ROWE JR., C. A. *J. Amer. chem. Soc.* 87 (1965) 3244; BANK, S. *ibid.* 87 (1965) 3245

148 KUHN, R. and REWICKI, D. *Angew. Chem.* 79 (1967) 648

149 BOWDEN, K. and COCKERILL, A. F. *Chemy. Comm.* (1967) 989

150 VOGEL, E. *et al. Angew. Chem.* 78 (1966) 643; RADLICK, P. and ROSEN, W. *J. Amer. chem. Soc.* 88 (1966) 3462

151 BAUM, G., BERNARD, R. and SHECHTER, H. *J. Amer. chem. Soc.* 89 (1967) 5307

152 WEBSTER, O. W. *J. Amer. chem. Soc.* 88 (1966) 3046

153 WEBSTER, O. W. *J. Amer. chem. Soc.* 88 (1966) 4055

154 RITCHIE, C. D. and USCHOLD, R. E. *J. Amer. chem. Soc.* 89 (1967) 2960; BRAUMAN, J. I., McMILLEN, D. F. and KONAZAWA, Y. *ibid.* 89 (1967) 1728; HOGEN-ESCH, T. E. and SMID, J. *ibid.* 89 (1967) 2764

155 EBEL, H. F. and WAGNER, B. O. unpublished work.

INDEX

2-Acetyl-2-decarboxamidotetracycline, 4
Acid strength,
  acidic hydrocarbons, pK-values, 131, 132, 133
  Brönsted relationship, 117, 118, 134
  calculation of intrinsic acidity, 122–126
  cyanocarbon acids, 133
  experimental, relation to HMO energies, 126; see also 144
  hydrocarbons, dependence on structure, 121–128
  hydrocarbons, measurement, 128–142
  thermodynamic acidity scales, 118–121
Acyl phosphates, as phosphorylating agents, 85, 86
N-Acylphosphoramidates as phosphorylating agents, 85
Adenosine triphosphate,
  synthesis from diphosphate, involving isoprenoid quinones, 103–105
  synthesis from diphosphate, involving nicotinamide-adenine dinucleotide, 99–103
  synthesis involving cytochromes a (ferroporphyrins), 108–111
Alkoxyvinyl phosphates as phosphorylating agents, 88
Allenes, from β-chloroalk-2-enyl phosphonic acids, 84
Allylphenacylammonium ions, electrophilic rearrangement, 50
Allylvinylammonium ions, electrophilic rearrangement, 68
Amidines, electrophilic rearrangement, 67
Amine oxides, electrophilic rearrangement, 62, 63
9-Aminofluorene, derivatives, electrophilic rearrangement, 58
Anhydroaureomycin nitrile, 6
Anhydrotetracycline, 5, 16
  chlorination, 15
  oxidation to tetracycline, 14
Anhydrotetracyclines,
  biosynthetic conversion into 5-hydroxytetracyclines, 22
  biosynthetic oxidation, 21
  biosynthetic route from pretetramids, 21
  photo-oxidation, 13
Aureomycin,
  dehydration, 6
  stereochemistry, 3

Benzhydryl ethers, electrophilic rearrangement, 56
Benzhydryltrimethylammonium ions, electrophilic rearrangement, 51, 52
Benzyl methyl ether, electrophilic rearrangements, 55, 56

N-Benzylnicotinamide chloride, use in phosphoryl transfer, 99
Benzylphenacylammonium ions, electrophilic rearrangements, 49, 50
Benzyl phenacyl ether,
  action of sodium butoxide, 56
Benzyl phenyl ether-
  action of phenyl-lithium, 56
Benzyltrimethylammonium ion, electrophilic rearrangement, 52
Benzyne,
  reaction with allyldiethylamine, 68
  reaction with sulphides, 61
  reaction with tertiary amines, 68
Biosynthesis,
  of salamander alkaloids, 45, 46
  of tetracyclines, 18–24
9-Bromo-anhydrotetracycline, 17
7-Bromotetracycline, 4

Carbanions, rearrangements in, 49–62, 68
Carbodiimides, as phosphorylating agents, 89
Catalytic hydroxylation of 12a-deoxytetracyclines, 16
Chloranil reaction with triethylamine, 97, 98
β-Chloroalk-2-enyl phosphonic acids, allenes from, 84
7-Chloro-5a,6-anhydrotetracycline, photo-oxidation, 13
2-Chlorodecylphosphonic acid, 83
7-Chloro-5,5a-dehydrotetracycline, 9
7-Chloro-5a,11a-dehydrotetracycline, 20
Chlorotetracyclines, biosynthesis, 20, 23, 24
N-Chlorsuccinimide, reaction with tetracyclines, 15
Cholesterol, role in biosynthesis of salamander alkaloids, 45, 46
Cinnamyl ethers, electrophilic rearrangement, 57
2-Crotyloxyquinoline, thermal rearrangement, 67
Cyanocarbon acids,
  pK's relative to water, 138–142
  comparison with calculation, 147
  reactions, 151
  relative pK-values, 133
  synthesis, 150, 151
9-Cyanofluorene, 147
  as standard for acidic hydrocarbons, 136, 137
Cycloneosamandaridine, 36, 44
Cycloneosamandione, 36, 44
  isolation, 35
  stereochemistry, 44, 45; see also 47
  structure, 43, 44
Cyclononatetraene salts, synthesis, 150

Cyclopropylmethylammonium ion,
electrophilic rearrangement, 52–54, 67
Cytochrome *a*,
role in phosphorylation, 108
structure, 109, 110
Cytochrome *b*, 100

(±)-Dedimethylamino-5a,6-anhydro-7-
chlorotetracycline, synthesis, 26
(±)-Dedimethylamino-decarboxamido-10,
12-dideoxytetracycline, synthesis, 30
(±)-Dedimethylamino-6-demethyl-6,12a-
dideoxy-7-chlorotetracycline, synthesis, 29
5a,11a-Dehydro-7-chlorotetracycline, 5
Dehydrotetracycline, 22
N-Demethylanhydrotetracycline, biosynthetic route from methylpretetramid,
23
6-Demethyl-6-deoxytetracycline,
electrophilic substitution, 9, 10
preparation, 9
6-Demethyltetracycline, 3, 8
12a-Deoxyanhydrotetracycline, catalytic
hydroxylation, 16
(±)-6-Deoxy-6-demethyltetracycline, synthesis, 26, 27
6-Deoxytetracyclines, 9
11a-fluoro derivatives, 12
12a-Deoxytetracyclines, hydroxylation at
C$_{12a}$, 16
12a-Desoxytetracyclines, bromination, 17
Di- and Trinitrophenyl phosphates, as
phosphorylating agents, 86
Dibenzylaniline, electrophilic rearrangement, 58
Dibenzylethyl ether, electrophilic rearrangement, 55, 56
Dibenzyldimethylammonium ion, electrophilic rearrangement, 52
2,3-Dichloro-5,6-dicyano-*p*-benzoquinone,
promoter in phosphoryl transfer, 95, 96
6,7-Dichloro-5,8-quinoline, promoter in phosphoryl transfer, 94, 95
Dicyclohexylcarbodiimide, 79
1,2-Dimethyl-5,6-cyclopentenonaphthalene,
from dehydrogenation of samandiol, 38
synthesis, 39
N,N-Dimethyl formamide, use in phosphoryl
transfer, 89, 90
2,2-Diphenylpropyl metallic derivatives,
electrophilic rearrangement, 62
N,N-Disubstituted aminomethylene compounds, in synthesis of acidic hydrocarbons, 148

Electron spin resonance spectra, of radicals
from acidic hydrocarbons, 144
Electrophilic rearrangements,
allylphenacylammonium ions, 51
amine oxides, 63, 64

ammonium nitrogen to anionic carbon,
49–55
ammonium nitrogen to anionic oxygen,
63, 64
benzhydryltrimethylammonium ions, 52
benzylphenacylammonium ions, 49, 50
benzyltrimethylammonium ion, 52, 53
definition, 50
elimination reactions, competition with,
58–60
ethers, of, 55–58
fluorenylammonium ions, 51
imino ethers and amidines, 67
1→2 migrations, 49–64
structural features favouring, 58–60
1→4-migrations, 64–66
'onium sulphur to anionic nitrogen, 62
oxygen to anionic carbon, 55–57
sulphonium ions to anionic carbon, 54
sulphur to anionic carbon, 61
uncharged carbon to anionic carbon, 61,
62
uncharged nitrogen to anionic carbon,
58, 59
Enol phosphates, as phosphorylating agents,
88
Epimerization,
at C$_4$ of tetracyclines, 6
at C$_6$ of tetracyclines, 9
1-Ethoxyethyl phosphate, as phosphorylating agent, 91

Ferro-complexes, structure, 108, 109
Ferro-porphyrins, structure and role in
phosphoryl transfer, 108–111
Fluoranil, promoter in phosphoryl transfer,
96, 97; *see also* 98
Fluorenylammonium ions, electrophilic rearrangement, 51; *see also* 52
Fluorenyl ethers, electrophilic rearrangements, 55

α-D-glucose-1 phosphate, as phosphorylating
agent, 89

β-Halogenoalkylphosphonates, as phosphorylating agents, 81
7-Halotetracyclines, reduction, 15
Hydrocarbon acids, 116–161
association of metal in solution, 129
calculation of intrinsic acidity, 122–126
carbanion reactivity, 153
formation of carbanions, 152
intrinsic acidity and thermodynamic
acidity scales, 118–121
isomerization, 155
measurement of acidity, 128–142
by equilibrium methods, 128
by kinetic methods, 134
oxidation of carbanions, 153
pK's relative to water, 135–142
p$_w$K-values, comparison with calculations,
144–147
relative pK-values, 131–133
structure-acidity relationship, 121–128

Hydrocarbon acids,
   synthesis
      via methylene compounds, 148
      via pentyn-1,5-diols, 149
   tautomeric hydrocarbons, 145
   tautomerism of carbanions, 152–155
   ultraviolet spectra of anions, 142–144
Hydroquinone phosphate as phosphorylating agent, 82
5-Hydroxytetracycline,
   biosynthesis, 21, 22
   tracer studies with, 18

Imino-ethers, rearrangement, 67
Imidoyl phosphates, as phosphorylating agents, 89, 99, 102
Isoindolinium ion, electrophilic rearrangement, 53
Isoprenoid quinones, role in oxidative phosphorylation, 104–108
Isoxazolium salts, use in phosphoryl transfer, 86

β-Ketoalkylphosphonates, as phosphorylating agents, 84

Methylene compounds in synthesis of acidic hydrocarbons, 148
6-Methylene tetracyclines, 11
   addition of mercaptans, 13
   formation, 12
Methylpretetramid, 19, 21
   biosynthetic conversion into N-demethyl-anhydrotetracycline, 23
Mitochondrial oxidation, reaction path, 100, 101

Naphthacene derivatives, role in biosynthesis of tetracyclines, 18, 19
Nicotinamide-adenine dinucleotide, role in oxidative phosphorylation, 99–104
4-Nitroaniline, as standard for acidic hydrocarbons, 136, 137
p-Nitrobenzylphosphonic acid, as phosphorylating agent, 84
Nitrones, electrophilic rearrangement, 63, 64
N-Nitrosophosphoramidates, as phosphorylating agents, 85

Oxidative phosphorylation, 75
   by cytochromes *a* (ferro-porphyrins), 108–111
   by isoprenoid quinones, mechanism of action, 103–105
   role of nicotinamide-adenine dinucleotide, 99–104
   role of ubiquinones, 99, 103–107
   with hydroquinone phosphates, mechanism, 93, 94, 104–108
4-Oxotetracyclines, 7

Pentacyanocyclopentadienide, 151
1,3,10,11,12-Pentahydroxynaphthacene, 18

Pentyn-1,5-diols, in synthesis of acidic hydrocarbons, 149
Perchlorylfluoride, reaction with tetracyclines, 15
Perphosphates, as phosphorylating agents, 92
Perphosphoric acid as phosphorylating agent, 82
Phenacylbenzylaniline, electrophilic rearrangement, 58
9-Phenylfluorene as standard for acidic hydrocarbons, 136, 137
Phosphine methylenes, bonding in, 77
Phosphoramidates, as phosphorylating agents, 79
Phosphoramidic acid, π-bond order in, 78
N-Phosphorimidazoles as phosphorylating agents, 88
Phosphorochloridic acid, esters, 79, 80
N-Phosphorocreatinine as phosphorylating agent, 87
Phosphoroguanidates as phosphorylating agents, 81, 87
Phosphorohydrazidates as phosphorylating agents, 92
Phosphorylating agents,
   acyl phosphates, 85, 86
   N-acylphosphoramidates, 85
   alkoxyalkylphosphates, 91
   anhydrides, 79
   π-bond order of phosphorus-oxygen bonds, need for modification of, 76–78
   carbodiimides, 89
   comparison with carboxylating agents, 76
   di- and trinitrophenyl phosphates, 86, 87
   enol phosphates, 88; *see also* 109
   general structural types, 81–83
   β-halogenoalkylphosphonates, 83
   hydroquinone phosphates, 82, 93, 94; *see also* 104–108
   imidoyl phosphates, 89, 99, 102
   isoxazolium salts, use of, 86
   β-ketoalkylphosphonates, 84
   p-nitrobenzylphosphonic acid, 84
   N-nitrosophosphoramidates, 85
   perphosphates, 92
   perphosphoric acid, 82
   phosphoramidates, 79–81
   N-phosphorimidazoles, 88
   phosphoroguanidates, 81, 87
   phosphorohydrazidates, 92
   O-phosphoryloximes, 92
   N-phosphourethanes, 85
   α-pyridyl phosphates, 90; *see also* 103
Phosphoryl transfer,
   general mechanisms, 76, 78
   promotion by dichlorodicyanobenzoquinone, 95
   promotion by fluoranil, 96, 97; *see also* 98
   promotion by quinones, 94–100
   promotion by tetramethoxybenzoquinone, 98; *see also* 108
Phosphoryl transfer reactions, 75–115
O-Phosphoryloximes as phosphorylating agents, 92

N-Phosphourethanes, as phosphorylating agents, 85
Photo-oxidation of anhydrotetracyclines, 13
Pretetramid, 19, 21
Pyridine-*N*-oxides, 2-alkoxy, thermal rearrangement, 70
α-Pyridyl phosphates, as phosphorylating agents, 90; *see also* 103
Δ¹-Pyrroline-l-oxides, use in phosphoryl transfer, 99

Quinoline-l-oxide, use in phosphoryl transfer, 99
Quinones, promoters in phosphoryl transfer, 94–101
Quinuclidine derivative, thermal rearrangement, 70

Salamander alkaloids, 35
  biosynthesis, 45, 46
  isolation, 35
  with oxazolidine system, 37–43
  without oxazolidine system, 43–45
Samandaridine, 36
  isolation, 36
  partial synthesis, 42
  structure, 41
Samandarine,
  O-acetyl-, 41
  from samandarone, 40
  isolation, 36
  oxidation to samandarone, 37
  reduction with lithium aluminium hydride, 38
  structure, elucidation of, 37–45
  toxicity, 35
Samandarine-hydrobromide, X-ray analysis, 40
Samandarone,
  from samandarine, 37
  isolation, 36
  structure, 41
Samandenone, 36, 39
  structure, 43
Samandesone, 37
Samandinine, 43
Samandiol, 38
  dehydrogenation, 38
Samanine, 44
Sarin, as phosphorylating agent, 92
Smiles rearrangement, 49
  and related changes, 64
Solvation energies, calculation, 127
Sommelet rearrangements,
  benzhydryltrimethylammonium ion, 52
  benzyltrimethylammonium ion, 52
  dibenzyldimethylammonium ion, 51, 52
  sulphides, 61
  sulphonium ions, 55
Sulphides, electrophilic rearrangements, 61
Sulphilimines, thermal rearrangement, 62

Sulphinic esters, rearrangement, 64
Sulphones,
  from sulphinic esters, 64
  hydroxy, rearrangement, 64

Terramycin, stereochemical features, 2, 3
Terrarubein, 19
Tetracyano-cyclopentadienide, 151
Tetracyclines,
  2-acetyldecarboxamido-, 4
  anhydro-, 13
  5a,6-anhydro-, 13
  biosynthesis, 5, 6, 18–24
  7-bromo-, 4
  C_4 modified, 6
  C_5-alkoxy-, 9
  chemical reactivity, 6–18
  7-chloro-; *see* aureomycin
  5a,lla-dehydro-7-chloro-, 5
  6-demethyl-, 3, 8, 9
  6-deoxy-, 9–12, 13
  12a-deoxy-, bromination, 17
  12a-deoxyanhydro-, hydroxylation, 16
  epimerization at C_4, 6, 16
  epimerization at C_6, 9
  5-hydroxy; *see* terramycin, 2
  hydroxylation at C_{12a}-, 16
  6-methylene-, 12, 13
  modified C_4 dimethylamino group, with, 8
  modified carboxamide group, with, 6
  naturally occurring, 2–6
  4-oxo-, 7, 8
  photo-oxidation at C_6, 13
  reaction with perchlorylfluoride, 15
  reduction of 12a-hydroxyl-, 16
  removal of C_4-dimethylamino group, 7
  stereochemistry, 3
  tetracycloxides, 8, 11
  total synthesis, 24–30
Tetramethyl phosphonium cation, deuterium exchange in, 77
Tetramethoxy-*p*-benzoquinone, promoter of phosphoryl transfer, 99
Tricyanomethane, 150
Trimethyl sulphonium cations, deuterium exchange in, 77

Ubiquinones,
  promoters of phosphoryl transfer, 99
  role in mitochondrial oxidative phosphorylation, 101, 103–107
Ultraviolet spectra, anions of acidic hydrocarbons, 142–144

Vitamin K_2, role in respiratory oxidative phosphorylation, 103, 104

Wawzonek condensation in synthesis of acidic hydrocarbons, 149
Wittig reagents, bonding in, 77

166